Grounded Consequence for Defeasible Logic

"Antonelli applies some of the techniques developed in Kripke's approach to the paradoxes to generalize some of the most popular formalisms for non-monotonic reasoning, particularly default logic. The result is a complex and sophisticated theory that is technically solid and attractive from an intuitive standpoint." – John Horty, *Committee on Philosophy and the Sciences, University of Maryland, College Park*

This is a monograph on the foundations of defeasible logic, which explores the formal properties of everyday reasoning patterns whereby people jump to conclusions, reserving the right to retract them in the light of further information. Although technical in nature, the book contains sections that outline basic issues by means of intuitive and simple examples.

G. Aldo Antonelli is Professor of Logic and Philosophy of Science at the University of California, Irvine.

Grounded Consequence for Defeasible Logic

G. ALDO ANTONELLI
University of California, Irvine

CAMBRIDGE UNIVERSITY PRESS
Cambridge, New York, Melbourne, Madrid, Cape Town, Singapore, São Paulo

Cambridge University Press
40 West 20th Street, New York, NY 10011-4211, USA

www.cambridge.org
Information on this title: www.cambridge.org/9780521842051

First published 2005

Printed in the United States of America

A catalog record for this publication is available from the British Library.

Library of Congress Cataloging in Publication Data

Antonelli, G. Aldo.
Grounded consequence for defeasible logic / G. Aldo Antonelli.
p. cm.
Includes bibliographical references and index.
ISBN 0-521-84205-0 (hardback)
1. Defeasible reasoning. I. Title.
BC199.D38A58 2005
160 – dc22 2005003176

ISBN-13 978-0-521-84205-1 hardback
ISBN-10 0-521-84205-0 hardback

Contents

List of Figures

Foreword

Logic is an ancient discipline that, ever since its inception some 2,500 years ago, has been concerned with the analysis of patterns of valid reasoning. The beginnings of such a study can be traced back to Aristotle, who first developed the theory of the *syllogism* (an argument form involving predicates and quantifiers). The field was further developed by the Stoics, who singled out valid patterns of *propositional* argumentation (involving sentential connectives), and indeed flourished in ancient times and during the Middle Ages, when logic was regarded, together with grammar and rhetoric (the other two disciplines of the *trivium*), as the foundation of humanistic education. However, the modern conception of logic is only approximately 150 years old, having been initiated in England and Germany in the latter part of the nineteenth century with the work of George Boole (*An Investigation of the Laws of Thought*, 1854), Gottlob Frege (*Begriffsschrift*, 1879), and Richard Dedekind (*Was sind und was sollen die Zahlen?*, 1888). Thus modern symbolic logic is a relatively young discipline, at least compared with other formal or natural sciences that have a long tradition.

Throughout its long history, logic has always had a *prescriptive* as well as a *descriptive* component. As a descriptive discipline, logic aims to capture the arguments accepted as valid in everyday linguistic practice. But this aspect, although present throughout the history of the field, has taken up a position more in the background since the inception of the modern conception of logic, to the point that it has been argued that the descriptive component is no longer part of logic proper, but belongs to other disciplines (such as linguistics or psychology). Nowadays logic is, first and foremost, a prescriptive discipline, concerned with the identification, analysis, and justification of valid inference forms.

The articulation of logic as a prescriptive discipline is, ideally, a two-fold task. The articulation first requires the identification of a class of valid arguments. The class thus identified must have certain features: not just any class of arguments will do. For instance, it is reasonable to require that the logical validity of an argument depends on only its *logical form*. This amounts to requiring that the class of valid argument be closed under the relation "having the same logical form as," in that, if an argument is classified as valid, then so is any other argument of the same logical form. If this is the case, then such an identification clearly presupposes, and rests on, a notion of logical form.

The question of what constitutes a good theory of logical form lies oustide the scope of this book, and hence it is not pursued any further. We shall limit ourselves to the observation that one can achieve the desired closure conditions by requiring that the class of valid arguments be generated in some uniform way from some restricted set of principles. For instance, Aristotle's theory of the syllogism accomplishes this in a characteristically elegant fashion. It classifies: subject–predicate propositions on the basis of their forms into a small number of classes, and one then generates syllogisms by allowing the two premises and the conclusion to take all possible forms.

The second part of the task, however, is much harder. Once a class of arguments is identified, one naturally wants to know what it is that makes these arguments *valid*. In other words, to accomplish this second task, one needs a general theory of *logical consequence*, a theory that was not only unavailable to the ancients, but that would not be available until the appearance of modern symbolic logic – when an effort was undertaken to formalize and represent mathematical reasoning – and that would not be completely developed until the middle of the twentieth century. It is only with the development of the first general accounts of the notion of logical consequence through the work of Alfred Tarski (Der Wahrheitsbegriff in den formalisierten Sprachen, 1935) that modern symbolic logic reaches maturity.

One of the salient features of such an account is a property known as *monotony*, according to which the set of conclusions logically following from a given body of knowledge grows proportionally to the body of knowledge itself. In other words, once a given conclusion has been reached, it cannot be "undone" by the addition of any amount of further information. This is a desirable trait if the relation of logical consequence is to capture the essential features of rigorous mathematical reasoning, in which conclusions follow from premises with a special kind of necessity that cannot be voided by augmenting the facts from which they are derived. Mathematical conclusions follow *deductively* from the premises – they are, in a sense, already *contained in* the premises – and they *must* be true whenever the premises are.

There is, however, another kind of reasoning, more common in everyday life, in which conclusions are reached tentatively, only possibly to be retracted when new facts are learned. This kind of reasoning is *nonmonotonic*, or, as we also say, *defeasible*. In everyday reasoning, people jump to conclusions on the basis of partial information, reserving the right to preempt those conclusions when more complete information becomes available.

It turns out that this kind of reasoning is quite difficult to capture formally in a precise way, and efforts in this direction are relatively new, when compared with the long and successful history of the efforts aimed at formalizing deductive reasoning. The main impetus for the formalization of defeasible reasoning comes from the artificial intelligence community, in which people realized very early on that everyday commonsense inferences cannot quite be represented in the golden standard of modern deductive logic, the first-order predicate calculus. Over the past two or three decades, a number of formalisms have been proposed to capture precisely this kind of reasoning, in an effort that has surpassed the boundaries of artificial intelligence proper, to become a new field of formal inquiry – nonmonotonic logic.

This book aims to contribute to this development by proposing an approach to defeasible reasoning that is in part inspired by parallel developments in philosophical logic, and in particular in the formal theory of truth. The point of view adopted here is the one just mentioned, that the formal study of defeasible reasoning – nonmonotonic logic – has come into its own as a separate field. Accordingly, the emphasis is more on conceptual, foundational issues and less on issues of implementation and computational complexity. (This is not to underestimate the salience of these topics, in fact they are mentioned whenever relevant – they just happen to fall outside the purvue of the book.)

The book is organized as follows. Chapter 1 starts focusing on the development of modern symbolic logic from the point of view of the abstract notion of logical consequence; in particular, we consider those features of logical consequence that aim to capture patterns of defeasible reasoning in which conclusions are drawn tentatively, subject to being retracted in the light of additional evidence. A number of useful nonmonotonic formalisms are briefly presented, with special emphasis on the question of obtaining well-behaved consequence relations for them. Further, a number of issues that arise in defeasible reasoning are treated, including skeptical versus creduluous reasoning, the special nature of so-called *floating conclusions*, and the conceptual distinction between an approach to the *nature* of conflict and the concrete question of how conflicts should be *handled*.

Chapter 2 deals with the problem of developing a direct approach to nonmonotonic inheritance over *cyclic* networks. This affords us the opportunity to develop the main ideas behind the present approach in the somewhat simpler setting of defeasible networks. The main thrust of the chapter is toward developing a notion of *general extension* for defeasible networks that not only applies to cyclic as well as acyclic networks, but also gives a *directly skeptical* approach to nonmonotonic inheritance.

Finally, Chaps. 3 and 4 further develop the approach by extending it to the much richer formalism of default logic. Here, the framework of general extensions is applied to Reiter's *default logic*, resulting in a well-behaved relation of defeasible consequence that vindicates the intuitions of the directly skeptical approach.

ACKNOWLEDGMENTS

I thank Rich Thomason and Jeff Horty for first introducing me to the field of defeasible reasoning, and Mel Fitting and Rohit Parikh for illuminating conversations on this topic. I am indebted to several anonymous referees for the *Artificial Intelligence Journal* and Cambridge University Press for providing useful feedback and criticisms. The critical example 3.5.2 in Chap. 3 was suggested by one of them, and its interpretation in the block world is due to Madison Williams.

I am grateful to Elsevier Publishing for permission to use material in Chaps. 2, 3, and 4 which first appeared in the *Artificial Intelligence Journal* (see Antonelli, 1997, 1999); and to Blackwell for using material in Chap. 1, which first appeared in Floridi (2004) (see Antonelli, 2004).

1

The Logic of Defeasible Inference

1.1 FIRST-ORDER LOGIC

It was mentioned that first-order logic (henceforth FOL) was originally developed for the representation of mathematical reasoning. Such a representation required the establishment of a high standard of rigor, meant to guarantee that the conclusion follows from the premises with absolute deductive cogency. In this respect, FOL turned out to be nothing but a stunning success. The account of deductive reasoning provided by FOL enjoys a number of important mathematical properties, which can also be used as a crucial benchmark for the assessment of alternative logical frameworks. (The reader interested in an introduction to the nuts and bolts of FOL can consult any of the many excellent introductory texts that are available, such as Enderton, 1972.)

From the point of view of abstract consequence relations, FOL provides an implementation of the so-called *no-counterexample* account: A sentence ϕ is a consequence of a set Γ of sentences if and only if one cannot reinterpret the (nonlogical part of the) language in which Γ and ϕ are formulated in such a way as to make all sentences in Γ true and ϕ false. An inference from premises $\psi_1, \ldots, \psi_k$ to a conclusion ϕ is *valid* if ϕ is a consequence of $\{\psi_1, \ldots, \psi_k\}$, i.e., if the inference has no counterexample.

For this to be a rigorous account of logical consequence, the underlying notion of interpretation needs to be made precise, along with a (noncircular, possibly stipulative) demarcation of the logical and nonlogical vocabulary. This was accomplished by Alfred Tarski in 1935, who precisely defined the notion of truth of a sentence on an interpretation (see Tarski, 1956, for a collection of Tarski's technical papers). In so doing,

Tarski overcame both a technical and a philosophical problem. The technical problem had to do with the fact that in FOL quantified sentences are obtained from components that are not, in turn, sentences, so that a direct recursive definition of truth for sentences breaks down at the quantifier case. To overcome this problem Tarski introduced the auxiliary notion of *satisfaction*. The philosophical obstacle had to do with the fact that the notion of *truth* was at the time considered suspiciously metaphysical among logicians trained within the environment of the Vienna Circle. This was a factor, for instance, in Gödel's reluctance to formulate his famous undecidability results in terms of truth (see, for instance, Feferman, 1998).

Tarski's analysis yielded a mathematically precise definition for the no-counterexample consequence relation of FOL, which is usually denoted by the symbol "$\models$": We say that ϕ is a consequence of a set Γ of sentences, written $\Gamma \models \phi$, if and only if ϕ is true on every interpretation on which every sentence in Γ is true. At first glance, there would appear to be something intrinsically infinitary about $\models$. Regardless of whether Γ is finite or infinite, to check whether $\Gamma \models \phi$ one has to "survey" infinitely many possible interpretations and check whether any of them is a counterexample to the entailment claim, i.e., whether any of them is such that all sentences in Γ are true on it while ϕ is false.

However, surprisingly, in FOL the infinitary nature of $\models$ is only apparent. As Gödel (1930) showed, the relation $\models$, although defined by universally quantifying over all possible interpretations, can be analyzed in terms of the existence of finite objects of a certain kind, viz., formal proofs. A *formal proof* is a finite sequence of sentences, each of which is an *axiom*, an *assumption*, or is obtained from previous ones by means of one of a finite number of inference rules, such as *modus ponens*. If a sentence ϕ occurs as the last line of a proof, then we say that the proof is a *proof of* ϕ, and we say that ϕ is *provable from* Γ, written $\Gamma \vdash \phi$, if and only if there is a proof of ϕ all of whose assumptions are drawn from Γ. In practice, in FOL, one provides a small and clearly defined number of primitive inferential principles (such as axioms and rules) and then posits that a conclusion ϕ follows from a set Γ of premises if ϕ can be obtained from some of the premises by repeated application of the inferential principles. Many different axiomatizations of FOL exist, and a particularly simple and elegant one can be found in Enderton (1972).

Gödel's famous completeness theorem of 1930 states that the two relations, $\models$ and $\vdash$, are extensionally equivalent: For any ϕ and Γ, $\Gamma \models \phi$ if and only if $\Gamma \vdash \phi$. This is a remarkable feature of FOL, which has a number of consequences. One of the deepest consequences follows from the fact that proofs are finite objects, and hence that $\Gamma \vdash \phi$ if and only if there is a *finite* subset Γ_0 of Γ such that $\Gamma_0 \vdash \phi$. This, together with the completeness theorem, gives us the *compactness theorem*: $\Gamma \models \phi$ if and only if there is a finite subset Γ_0 of Γ such that $\Gamma_0 \models \phi$. There are many interesting equivalent formulations of the theorem, but the following one is perhaps the most often cited. Say that a set of sentences is *consistent* if they can all be made simultaneously true on some interpretation; then the compactness theorem says that a set Γ is consistent if and only if each of its finite subsets is by itself consistent.

Another important consequence of Gödel's completeness theorem is the following form of the Löwenheim–Skolem theorem: If all the sentences in Γ can be made simultaneously true on some interpretation, then they can also be made simultaneously true on some (other) interpretation whose universe is no larger than the set $\mathbb{N}$ of the natural numbers.

Together, the compactness and the Löwenheim–Skolem theorems are the beginning of one of the most successful branches of modern symbolic logic: model theory. The compactness and the Löwenheim–Skolem theorems characterize FOL; as shown by Per Lindström in 1969, any logical system (meeting certain "regularity" conditions) for which both compactness and Löwenheim–Skolem hold is no more expressive than FOL (see Ebbinghaus, Flum, and Thomas, 1994, Chap. XIII, for an accessible treatment).

Gödel's completeness theorem also reflects on the question of whether and to what extent one can devise an effective procedure to determine whether a sentence ϕ is valid or, more generally, if $\Gamma \models \phi$ for given Γ and ϕ. First, some terminology. We say that a set Γ of sentences is *decidable* if there is an effective procedure, i.e., a mechanically executable set of instructions that determines, for each sentence ϕ, whether ϕ belongs to Γ or not. Notice that such a procedure gives both a positive and a negative test for membership of a sentence ϕ in Γ. A set of sentences is *semidecidable* if there is an effective procedure that determines if a sentence ϕ is a member of Γ, but might not provide an answer in some cases in which ϕ is not a member of Γ. In other words, Γ is semidecidable if there is a positive, but not necessarily a negative, test for membership in Γ. Equivalently, Γ is semidecidable if it can be given an effective listing, i.e., if it can be mechanically generated. These notions can be generalized to relations among sentences of any number of arguments. For instance, it is an important feature of the axiomatizations of FOL, such as that of Enderton (1972), that both the set of axioms and the relation that holds among $\phi_1, \ldots, \phi_k$ and ψ when ψ can be inferred from $\phi_1, \ldots, \phi_k$ by one application of the rules, are decidable. As a result, the relation that holds among $\phi_1, \ldots, \phi_k$ and ϕ whenever $\phi_1, \ldots, \phi_k$ is a proof of ϕ is also decidable.

The import of Gödel's completeness theorem is that if the set Γ is decidable (or even only semidecidable), then the set of all sentences ϕ such that $\Gamma \models \phi$ is semidecidable. Indeed, one can obtain an effective listing for such a set by systematically generating all proofs from Γ. The question arises of whether, in addition to this positive test, there might not be a negative test for a sentence ϕ being a consequence of Γ. This *decision problem* [*Entscheidungsproblem*] was originally proposed by David Hilbert in 1900, and it was solved in 1936 independently by Alonzo Church and Alan Turing. The Church–Turing theorem states that, in general, it is not decidable whether $\Gamma \models \phi$, or even whether ϕ is valid. (It's important to know that for many, even quite expressive, fragments of FOL the decision problem is solvable; see Börger, Grädel, and Gurevich, 1997, for details.) We should also notice the following fact that will be relevant in Section 1.3; say that a sentence ϕ is *consistent* if $\{\phi\}$ is consistent, i.e., if its negation $\neg\phi$ is not valid. Then the set of all sentences ϕ such that ϕ is consistent is not even semidecidable, for a positive test for such a set would yield a negative test for the set of all valid sentences, which would so be decidable, against the Church–Turing theorem.

1.2 CONSEQUENCE RELATIONS

In the previous section, we considered the no-counterexample consequence relation $\models$ by saying that $\Gamma \models \phi$ if and only if ϕ is true on every interpretation on which every sentence in Γ is true. In general, it is possible to consider the abstract properties of a relation of consequence between sets of sentences and single sentences. Let $\mid\!\sim$ be any such relation. We identify the following properties, all of which are satisfied by the consequence relation $\models$ of FOL:

Supraclassicality: If $\Gamma \models \phi$ then $\Gamma \mid\!\sim \phi$;
Reflexivity: If $\phi \in \Gamma$ then $\Gamma \mid\!\sim \phi$;
Cut: If $\Gamma \mid\!\sim \phi$ and $\Gamma, \phi \mid\!\sim \psi$ then $\Gamma \mid\!\sim \psi$;
Monotony: If $\Gamma \mid\!\sim \phi$ and $\Gamma \subseteq \Delta$ then $\Delta \mid\!\sim \phi$.

Supraclassicality states that if ϕ follows from Γ in FOL, then it also follows according to $\mid\!\sim$; i.e., $\mid\!\sim$ extends $\models$ (the relation $\models$ is trivially supraclassical). Of the remaining conditions, the most straightforward is Reflexivity: It says that if ϕ belongs to the set Γ, then ϕ is a consequence of Γ. This is a very minimal requirement on a relation of logical consequence. We certainly would like all sentences in Γ to be inferable from Γ. It's not clear in what sense a relation that fails to satisfy this requirement can be called a *consequence* relation.

Cut, a form of transitivity, is another crucial feature of consequence relations. Cut is as a conservativity principle: If ϕ is a consequence of Γ, then ψ is a consequence of Γ together with ϕ only if it is already a consequence of Γ alone. In other words, adjoining to Γ something that is already a consequence of Γ does not lead to any *increase* in inferential power. Cut can be regarded as the statement that the "length" of a proof does not affect the degree to which the assumptions support the conclusion. Where ϕ is already a consequence of Γ, if ψ can be inferred from Γ together with ϕ, then ψ can also be obtained by means of a longer "proof" that proceeds indirectly by first inferring ϕ. It is immediate to check that FOL satisfies Cut.

It is worth noting that many forms of probabilistic reasoning fail to satisfy Cut, precisely because the degree to which the premises support the conclusion is inversely correlated to the length of the proof. To see this, we adapt a well-known example. Let Ax stand for "x was born in Pennsylvania Dutch country," Bx stand for "x is a native speaker of German," and Cx stand for "x was born in Germany." Further, let Γ comprise the statements "most As are Bs," "most Bs are Cs," and Ax. Statements of the form "most As are Bs" are interpreted probabilistically as saying that the conditional probability of B given A is, say, greater than 50%; likewise, we say that Γ supports a statement ϕ if Γ assigns ϕ a probability $p > 50\%$.

Then Γ supports Bx, and Γ together with Bx supports Cx, but Γ by itself does not support Cx. Because Γ contains "most As are Bs" and Ax, it supports Bx (in the sense that the probability of Bx is greater than 50%); similarly, Γ together with Bx supports Cx; but Γ by itself cannot support Cx. Indeed, the probability of someone who was born in Pennsylvania Dutch country being born in Germany is arbitrarily close to zero. Examples of inductive reasoning such as the one just given cast some doubt on the possibility of coming up with a logically well-behaved relation of probabilistic consequence.

Special considerations apply to Monotony. Monotony states that if ϕ is a consequence of Γ then it is also a consequence of any set containing Γ (as a subset). The import of Monotony is that one cannot preempt conclusions by adding new premises to the inference. It is clear why FOL satisfies Monotony: Semantically, if ϕ is true on every interpretation on which all sentences of Γ are true, then ϕ is also true on every interpretation on which all sentences in a larger set Δ are true (similarly, proof theoretically, if there is a proof of ϕ, all of whose assumptions are drawn from Γ, then there is also a proof of ϕ – indeed, the same proof – all of whose assumptions are drawn from Δ).

Many people consider this feature of FOL as inadequate to capture a whole class of inferences typical of everyday (as opposed to mathematical or formal) reasoning and therefore question the descriptive adequacy of FOL when it comes to representing commonsense inferences. In everyday life, we quite often reach conclusions tentatively, only to retract them in the light of further information. Here are some typical examples of essentially nonmonotonic reasoning patterns.

TAXONOMIES. Taxonomic knowledge is essentially hierarchical, with superclasses subsuming smaller ones: Poodles are dogs, and dogs are mammals. In general, subclasses inherit features from superclasses: All mammals have lungs, and because dogs are mammals, dogs have lungs as well. However, taxonomic knowledge is seldom strict, in that feature inheritance is prone to exceptions: Birds fly, but penguins (a special kind of bird) are an exception. Similarly, mammals don't fly, but bats (a special kind of mammal) are an exception.

It would be unwieldy (to say the least) to provide an exhaustive listing of all the exceptions for each subclass–superclass pair. It is therefore natural to interpret inheritance *defeasibly*, on the assumption that subclasses inherit features from their superclasses, unless this is explicitly blocked. For instance, when told that Stellaluna is a mammal, we infer that she does not fly, because mammals, by and large, don't fly. But the conclusion that Stellaluna doesn't fly can be retracted when we learn that Stellaluna is a bat, because bats are a specific kind of mammal, and they do fly. So we infer that Stellaluna does fly after all. This process can be further iterated. We can learn, for instance, that Stellaluna is a baby bat and that therefore she does not know how to fly yet. Such complex patterns of defeasible reasoning are beyond the reach of FOL, which is, by its very nature, monotonic.

CLOSED WORLD. Some of the earliest examples motivating defeasible inference come from database theory. Suppose you want to travel from Oshkosh to Minsk and therefore talk with your travel agent who, after querying the airline database, informs you that there are no direct flights. The travel agent doesn't actually *know* this, as the airline database contains explicit information only about existing flights. However, the database incorporates a *closed-world assumption* to the effect that the database is complete. But the conclusion that there are no direct connections between Oshkosh and Minsk is defeasible, as it could be retracted on expansion of the database.

DIAGNOSTICS. When complex devices fail, it is reasonable to assume that the failure of a smallest set of components is responsible for the observed behavior. If the failure of any two out of three components A, B, and C, can explain the device's failure, it is assumed that not all three components simultaneously fail, an assumption that can be retracted in the light of further information (e.g., if replacement of A and B fails to restore the expected performance).

For these and similar reasons, people have striven, over the past 25 years or so, to devise nonmonotonic formalisms capable of representing defeasible inference. We will take a closer look at these formalisms in Section 1.3, but for now we want to consider the issue from a more abstract point of view.

When one gives up Monotony in favor of descriptive adequacy, the question arises of what formal properties of the consequence relation are to take the place of Monotony. Two such properties have been considered in the literature for an arbitrary consequence relation $\mid\sim$:

Cautious Monotony: If $\Gamma \mid\sim \phi$ and $\Gamma \mid\sim \psi$, then $\Gamma, \phi \mid\sim \psi$;
Rational Monotony: If $\Gamma \not\mid\sim \neg\phi$ and $\Gamma \mid\sim \psi$, then $\Gamma, \phi \mid\sim \psi$.

Both Cautious Monotony and the stronger principle of Rational Monotony are special cases of Monotony and are therefore not in the foreground as long as we restrict ourselves to the classical consequence relation $\models$ of FOL.

Although superficially similar, these principles are quite different. Cautious Monotony is the converse of Cut: It states that adding a consequence ϕ back into the premise set Γ does not lead to any *decrease* in inferential power. Cautious Monotony tells us that inference is a cumulative enterprise: We can keep drawing consequences that can in turn be used as additional premises, without affecting the set of conclusions. Together with Cut, Cautious Monotony says that if ϕ is a consequence of Γ then for any proposition ψ, ψ is a consequence of Γ if and only if it is a consequence of Γ together with ϕ. In other words, as pointed out by Kraus, Lehman, and Magidor (1990, p. 178), if the new facts turned out already to be expected to be true, nothing should change in our belief system. It also turns out that Cautious Monotony has a nice semantic characterization: The just-cited article by Kraus et al. (1990) provides a system **C** (with Cautious Monotony among its axioms), which is proved sound and complete with respect to entailment over suitably defined *preferential models*, having a preferential ordering $\prec$ between states. In fact, it has been often pointed out that Reflexivity, Cut, and Cautious Monotony are critical properties for any well-behaved nonmonotonic consequence relation (see Gabbay, Hogger, and Robinson, 1994; Stalnaker, 1994).

The status of Rational Monotony is much more problematic. As we observed, Rational Monotony can be regarded as a strengthening of Cautious Monotony, and, like the latter, it is a special case of Monotony. A case for Rational Monotony is forcefully made in Lehman and Magidor (1992, p. 20), as follows. Let p, q, and r be distinct propositional variables, and suppose that $p \mid\!\sim q$ (for instance, because it is explicitly contained in our knowledge base); then we would intuitively expect also $p, r \mid\!\sim q$, as r cannot possibly provide any information about whether p is satisfied or not (and in particular $p \not\mid\!\sim \neg r$). Observe that there are relevance considerations at work here. The reason that $p, r \mid\!\sim q$ appears plausible to us is that the sentences involved are atomic and therefore none of them is relevant for the truth of any of the others.

We will come back to this issue of relevance in Section 1.6, but for now we observe that there are reasons to think that Rational Monotony might not be a correct feature of a nonmonotonic consequence relation after all. Stalnaker (1994, p. 19) adapts a counterexample drawn from the literature on conditionals. Consider three composers: Verdi, Bizet, and Satie. Suppose that we initially accept (correctly but defeasibly) that Verdi is Italian, whereas Bizet and Satie are French. Suppose now that we are told by a reliable source of information that Verdi and Bizet are compatriots. This leads us no longer to endorse the propositions that Verdi is Italian (because he could be French), and that Bizet is French (because he could be Italian); but we would still draw the defeasible consequence that Satie is French, because nothing that we have learned conflicts with it. By letting $I(v)$, $F(b)$, and $F(s)$ represent our initial beliefs about the nationality of the three composers, and $C(v, b)$ represent that Verdi and Bizet are compatriots, the situation could be represented as follows:

$$C(v, b) \mid\!\sim F(s).$$

Now consider the proposition $C(v, s)$ that Verdi and Satie are compatriots. Before learning that $C(v, b)$ we would be inclined to reject the proposition $C(v, s)$ because we endorse $I(v)$ and $F(s)$, but after learning that Verdi and Bizet are compatriots, we can no longer endorse $I(v)$, and therefore we no longer reject $C(v, s)$. The situation then is as follows:

$$C(v, b) \not\mid\!\sim \neg C(v, s).$$

However, if we added $C(v, s)$ to our stock of beliefs, we would lose the inference to $F(s)$: In the context of $C(v, b)$, the proposition $C(v, s)$ is equivalent to the statement that all three composers have the same nationality, and this leads us to suspend our assent to the proposition $F(s)$. In other words, and contrary to Rational Monotony,

$$C(v, b), C(v, s) \not\mid\!\sim F(s).$$

Thus we have a counterexample to Rational Monotony. On the other hand, there appear to be no reasons to reject Cautious Monotony, which is in fact a characteristic feature of our reasoning process. In this way we come to identify four crucial properties of a nonmonotonic consequence relation: Supraclassicality, Reflexivity, Cut, and Cautious Monotony.

1.3 NONMONOTONIC LOGICS

As was mentioned, over the past 25 years or so, a number of socalled *nonmonotonic* logical frameworks have emerged, expressly devised for the purpose of representing defeasible reasoning. The development of such frameworks represents one of the most significant developments both in logic and artificial intelligence and has wide-ranging consequences for our philosophical understanding of argumentation and inference.

Pioneering work in the field of nonmonotonic logics was carried out beginning in the late 1970s by (among others) J. McCarthy, D. McDermott, J. Doyle, and R. Reiter (see Ginsberg, 1987, for a collection of early papers in the field). With these efforts, the realization (which was hardly new) that ordinary FOL was inadequate to represent defeasible reasoning was for the first time accompanied by several proposals of formal frameworks within which one could at least begin to talk about defeasible inferences in a precise way, with the long-term goal of providing for defeasible reasoning an account that could at least approximate the degree of success achieved by FOL in the formalization of mathematical reasoning. The publication of a monographic issue of the *Artificial Intelligence Journal* in 1980 can be regarded as the "coming of age" of defeasible formalisms.

The development of nonmonotonic logics has been guided all along by a rich supply of examples. Many of these examples share the feature of an attempted *minimization* of the extension of a particular predicate (a minimization that is not, in general, representable in FOL, or at least not in a natural way). For instance, recall the travel agent example that was used in the preceding section in discussing the closed-world assumption: What we have in this example is an attempt to *minimize* the extension of the predicate "flight between." And, of course, such a minimization needs to take place not with respect to what the database explicitly contains but with respect to what it implies.

The idea of minimization is at the basis of one of the earliest nonmonotonic formalisms, McCarthy's *circumscription*. Circumscription makes explicit the intuition that, all other things being equal, extensions of predicates should be *minimal*. Again, consider principles such as "all normal birds fly." Here we are trying to minimize the extension of the abnormality predicate and assume that a given bird is normal unless we have positive information to the contrary. Formally, this can be represented using second-order logic. In second-order logic, in contrast to FOL, one is allowed to explicitly quantify over predicates, forming sentences such as $\exists P \forall x Px$ ("there is a universal predicate") or $\forall P(Pa \leftrightarrow Pb)$ ("a and b are indiscernible"). In circumscription, given predicates P and Q, we abbreviate $\forall x(Px \rightarrow Qx)$ ("all Ps are Qs") as $P \leq Q$, and likewise we abbreviate $P \leq Q \wedge Q \not\leq P$ as $P < Q$. If $A(P)$ is a formula containing occurrences of a predicate P, then the circumscription of P in A is the following second-order sentence $A^*(P)$:

$$A(P) \wedge \neg\exists Q[A(Q) \wedge Q < P].$$

$A^*(P)$ says that P satisfies A and that no smaller predicate does. Let Px be the predicate "x is abnormal," and let $A(P)$ be the sentence "all normal birds fly." Then the sentence "Tweety is a bird," together with $A^*(P)$ implies the sentence "Tweety flies," for the circumscription axiom forces the extension of P to be empty, so that "Tweety is normal" is automatically true. In terms of consequence relations, circumscription allows us to define, for each predicate P, a nonmonotonic relation $A(P) \mid\!\sim \phi$ that holds precisely when $A^*(P) \models \phi$. (This basic form of circumscription has been generalized, for in practice, one needs to minimize the extension of a predicate while allowing the extension of certain other predicates to vary.) From the point of view of applications, however, circumscription has a major shortcoming because of the second-order nature of $A^*(P)$. In general, second-order logic does not have a complete inference procedure: The price one pays for the greater expressive power of second-order logic is that there are no complete axiomatizations, as we have for FOL. It follows that it is impossible to determine whether $A(P) \mid\!\sim \phi$ [except in special cases in which $A^*(P)$ happens to be in fact equivalent to a first-order sentence (see Lifschitz, 1987)].

There is another family of approaches to defeasible reasoning that makes use of a *modal* apparatus, most notably *autoepistemic logics*. Modal logics in general have proved to be one of the most flexible tools for modeling many kinds of dynamic processes and their complex interactions. Besides the applications in knowledge representation, which are subsequently treated, there are modal frameworks, known as *dynamic logics*, that play a crucial role, for instance, in the modeling of serial or parallel computation. The basic idea of modal logic is that the language is interpreted with respect to a given set of *states* and that sentences are evaluated relative to one of these states. What these states are taken to represent depends on the particular application under consideration (they could be epistemic states or states in the evolution of a dynamical system, etc.), but the important thing is that there are *transitions* (of one or more different kinds) between states, and different modal logics are classified according to the properties of the associated transitions. For instance, in the case of one transition that is both *transitive* (i.e., such that if $a \rightarrow b$ and $b \rightarrow c$ then $a \rightarrow c$) and *euclidean* (if $a \rightarrow b$ and $a \rightarrow c$ then $b \rightarrow c$), the resulting modal system is referred to as K45. The different state transitions are formally represented in the language by distinct modalities, usually written as a box $\Box$. A sentence of the form $\Box A$ is true at a state s if and only if A is true at every state s' reachable from s by the kind of transition associated with $\Box$ (see Chellas, 1980, or Cresswell and Hughes, 1995, for comprehensive and accessible introductory treatments of modal logic).

In autoepistemic logic, the states involved are epistemic states of the agent (or agents). The intuition underlying autoepistemic logic is that we can sometimes draw inferences concerning the state of the world by using information concerning our own knowledge or ignorance. For instance, I can conclude that I do not have a sister given that if I did I would probably know about it, and nothing to that effect is present in my "knowledge base." But such a conclusion is defeasible, as there is always the possibility of learning new facts.

To make these intuitions precise, consider a modal language in which the necessity operator $\Box$ is interpreted as "it is known that." As with other defeasible formalisms, as we will see, the central notion in autoepistemic logic is that of an *extension* of a theory S, i.e., a consistent and self-supporting set of beliefs that can reasonably be entertained on the basis of S. Given a set S of sentences, let S_0 be the subset of S composed of those sentences containing no occurrences of $\Box$; further, let the *introspective closure* S_0^i of S_0 be the set

$$\{\Box\phi : \phi \in S_0\},$$

and the *negative introspective closure* S_0^n of S_0 be the set

$$\{\neg\Box\phi : \phi \notin S_0\}.$$

The set S_0^i is called the introspective closure because it explicitly contains positive information about the agent's epistemic state: S_0^i expresses what is known by the agent (and similarly, S_0^n contains negative information about the agent's epistemic status, stating explicitly what is not known). With these notions in place, we define an extension for S to be a set T of sentences such that

$$T = \{\phi : \phi \text{ follows from } S \cup T_0^i \cup T_0^n \text{ in K45}\}.$$

Autoepistemic logic gives us a rich language, with interesting mathematical properties and connections to other nonmonotonic formalisms, and provides a defeasible framework with well-understood modal properties.

Another nonmonotonic formalism that, like circumscription, is inspired by the intuition of minimization of abnormalities is *nonmonotonic inheritance*. We will deal with nonmonotonic inheritance (as well as with Default Logic) at length in this book (see Chaps. 2 and 3), but for now a brief introduction will suffice. As we have seen in Section 1.2, whenever we have a taxonomically organized body of knowledge, we presuppose that subclasses inherit properties from their superclasses. However, there can be exceptions, which can interact in complex ways. To use an example already introduced, mammals, by and large, don't fly; because bats are mammals, in the absence of any information to the contrary, we are justified in inferring that bats do not fly. Then we learn that bats are exceptional mammals, in that they do fly: The conclusion that they don't fly is retracted, and the conclusion that they fly is drawn instead. Things can be more complicated still, for in turn, as we have seen, baby bats are exceptional bats, in that they do not fly (does that make them unexceptional mammals?). Here we have potentially *conflicting inferences*. When we infer that Stellaluna, being a baby bat, does not fly, we are resolving all these potential conflicts based on a *specificity* principle: More specific information overrides more generic information. Nonmonotonic inheritance networks were developed for the purpose of capturing taxonomic examples such as the preceding one. Such networks are collections of nodes and directed ("is a") links representing taxonomic information. When exceptions are allowed, the network is interpreted *defeasibly*. Figure 1.1 gives a network representing this state of affairs. In such a network, links of the form $A \rightarrow B$ represent the fact that typical As are Bs, and links $A \not\rightarrow B$

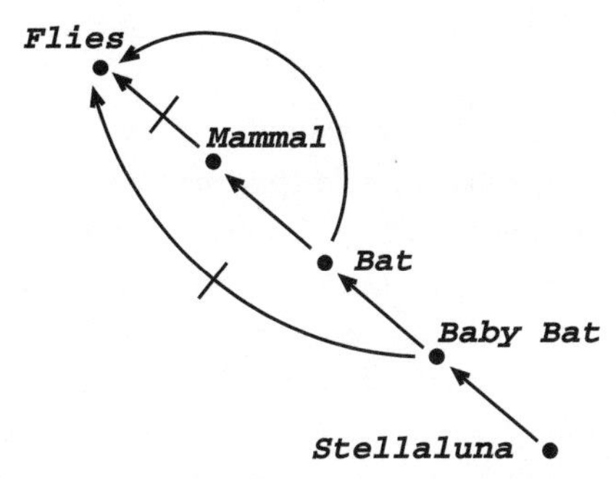

FIGURE 1.1. An inheritance network.

represent the fact that typical *A*s are not *B*s. If the network contains a link of the form $A \to B$, then information about *A*s is more specific than information about *B*s, and hence should override it. Research on nonmonotonic inheritance focuses on the different ways in which one can make this idea precise.

The main issue in defeasible inheritance is to characterize the set of assertions that are supported by a given network. It is of course not enough to devise a representational formalism; one also needs to specify how the formalism is to be interpreted, and this is precisely the focus of much work in nonmonotonic inheritance. Such a characterization is again accomplished through the notion of *extension* of a given network. There are two competing characterizations of extension for this kind of network, one that follows a strategy that has come to be known as *credulous*, and one that follows a strategy that has come to be known as *skeptical*. According to the credulous strategy, one always commits to as many assertions as possible, subject to the requirement that the set of endorsed assertions be free of conflict. According to the skeptical strategy, when two assertions are mutually conflicting, both are withheld.

Both strategies proceed by first defining the *degree* of a path through the network as the length of the longest sequence of links connecting its endpoints, and then building extensions by considering paths in ascending order of their degrees. The details are not reviewed here, as many of the same issues arise in connection with Default Logic (which is subsequently treated to greater length), but Horty (1994) provides an extensive survey. Because the notion of degree makes sense only in the case of acyclic networks, special issues arise when networks contain cycles (see

Chap. 2, based on Antonelli, 1997, for a treatment of inheritance on cyclic networks).

Although the language of nonmonotonic networks is expressively limited by design (in that only links of the form "is a" can be represented in a natural fashion), such networks represent an extremely useful setting in which to test and hone one's intuitions and methods for handling defeasible information, which can then be extended to more expressive formalisms. Among the latter is Reiter's *Default Logic*, which is perhaps the most flexible among nonmonotonic frameworks (again, see Chap. 3, based on Antonelli, 1999, for an extensive treatment). In Default Logic, the main representational tool is that of a *default rule*, or simply a *default*. A default is a *defeasible inference rule* of the form

$$\frac{\eta : \theta}{\xi}, \tag{1.1}$$

where η, θ, and ξ are sentences in a given language, respectively called the prerequisite, the justification, and the conclusion of the default. The interpretation of the default is that if η is known, and there is no evidence that θ might be false, then the rule allows the inference of ξ. As is clear, application of the rule requires that a consistency condition be satisfied, and rules can interact in complex ways. In particular it is possible that application of a rule might cause the consistency condition to fail (as when θ is $\neg\xi$). Reiter's Default Logic uses its own notion of an extension to make precise the idea that the consistency condition has to be met both before and after the rule is applied. Given a set Γ of defaults, an extension for Γ represents a set of inferences that one can reasonably and consistently draw by using defaults from Γ. More in particular (and in typical circular fashion), an extension for Γ comprises the conclusions of a maximal subset Δ of Γ, such that the conclusions of defaults in Δ both imply all the prerequisites of defaults in Δ and are consistent with all the justifications of defaults in Δ.

This definition can be made precise as follows. By a *default theory* we mean a pair (W, Δ), where Δ is a (finite) set of defaults and W is a set of sentences (a world description). The idea is that W represents the strict or background information, whereas Δ specifies the defeasible information. Given a pair (T_1, T_2) of sets of sentences, a default of the form of expression (1.1) is *triggered* by (T_1, T_2) if and only if $T_1 \models \eta$ and $T_2 \not\models \neg\theta$ (i.e., θ is consistent with T_2). Notice how this definition is built "on top" of $\models$: We could, conceivably, employ a different relation here (an issue that is considered again in Section 1.6). Finally we say that a set of sentences E is an extension for a default theory (W, Δ) if and only if

$$E = E_0 \cup E_1 \cup \ldots \cup E_n \cup \ldots,$$

where $E_0 = W$ and

$$E_{n+1} = E_n \cup \left\{ \xi : \frac{\eta : \theta}{\xi} \in \Delta \text{ is triggered by } (E_n, E) \right\}$$

(notice the occurrence of the limit E in the definition of E_{n+1}). There is an alternative characterization of extensions: Given a default theory, let $\mathfrak{S}$ be an operator defined on sets of sentences such that, for any set S of sentences, $\mathfrak{S}(S)$ is the smallest set containing W, deductively closed [i.e., such that if $\mathfrak{S}(S) \models \phi$ then $\phi \in \mathfrak{S}(S)$], and such that if a default with consequent ξ is triggered by (S, S) then $\xi \in \mathfrak{S}(S)$. Then one can show that E is an extension for (W, Δ) if and only if E is a fixed point of $\mathfrak{S}$, i.e., if $\mathfrak{S}(E) = E$.

For any given default theory, extensions need not exist, and even when they exist, they need not be unique. Let us consider a couple of examples. Our first example is a default theory that has no extension: Let W contain the sentence η, and let Δ comprise the single default:

$$\frac{\eta : \theta}{\neg\theta}.$$

If E were an extension, then the preceding default would have to be either triggered or not triggered by it, and either case is impossible. The default cannot be triggered, for then its justification would be inconsistent with the extension; but then the justification *is* consistent with E, so that the default is triggered after all.

This is a peculiar phenomenon, which bears a resemblance to the so-called *Liar Paradox*. This paradox, as is well known, arises when we consider the following sentence Λ: "Λ is not true." It is then impossible consistently to hold or deny that Λ is true. Philosophers and logicians have developed a number of solutions to the paradox, but one in particular is relevant here. Kripke (1975) shows that it is possible to provide semantics for a language containing its own truth predicate (and in which therefore Λ can be expressed), provided we give up *bivalence*; provided, that is, that we switch (for instance) to a three-valued setting. The desired semantics is then achieved by means of a fixpoint construction.

Once we allow a language to contain its own truth predicate, there are many other self-referential sentences that can be constructed, and not all as pathological as the Liar. For example, the following sentence T is known as the *Truth-Teller*: "T is true." There is a sense in which this sentence is as pathological as the Liar, in that it escapes the recursive clauses of a truth definition *à la* Tarski (see Gupta and Belnap, 1993, for an extensive discussion of truth and paradox). There is an important difference, however: Contrary to what happens for the Liar, it *is* possible consistently to hold that T is true, *and* it is possible consistently to deny that T is true (although not at the same time). That is, sentence T, contrary to the Liar, does not force us to renounce bivalence. The similarity between self-defeating defaults and three-valued approaches to the theory of truth is exploited in providing the notion of a *general extension* (for default theories and defeasible networks) in Chaps. 2 and 3.

Let us now consider an example of a default theory with multiple extensions. As before, let W contain the sentence η, and suppose Δ comprises the two defaults

$$\frac{\eta : \theta}{\neg\xi}, \qquad \frac{\eta : \xi}{\neg\theta}.$$

This theory has exactly two extensions, one in which the first default is triggered and one in which the second one is. It is easy to see that at least a default has to be triggered in any extension and that both defaults cannot be triggered by the same extension.

These examples are enough to bring out a number of features. First, it should be noted that neither one of the two characterizations of extensions for default logic just given gives us a way to "construct" extension by means of anything resembling an iterative process. Essentially, we have to "guess" a set of sentences E, and then verify that it satisfies the definition of an extension.

Further, the fact that default theories can have zero, one, or more extensions raises the issue of what inferences we are warranted in drawing from a given default theory. The problem can be presented as follows: Given a default theory (W, Δ), what sentences ϕ can be regarded as *defeasible consequences* of the theory? That is, for what ϕs does it hold that $(W, \Delta) \mid\!\sim \phi$? At first glance, there are several options available.

One option is to take the union of the extensions of the theory and consider ϕ a consequence of a default theory (W, Δ) if and only if $\phi \in E$, for some extension E. But this option needs to be ruled out, because it can sometimes lead to endorsing contradictory conclusions, as in the following example, in which $\alpha \in W$ and Δ comprises

$$\frac{\alpha : \beta}{\gamma}, \qquad \frac{\alpha : \beta}{\neg\gamma}.$$

This theory has two extensions, each triggering exactly one of the two defaults. Because the defaults have conflicting conclusions, if we defined $\mid\sim$ as we did previously we would have $(W, \Delta) \mid\sim \gamma$ as well as $(W, \Delta) \mid\sim \neg\gamma$, i.e., we would be led to endorse conflicting conclusions. In turn, this violates a property that many consider necessary for any viable notion of defeasible consequence for default logic, viz., the property that the set $\{\phi : (W, \Delta) \mid\sim \phi\}$ must be consistent whenever W is.

Once this first option is ruled out, only two alternatives are left, similar to the credulous and skeptical strategies for defeasible networks. On the "credulous" or "bold" strategy, one picks an extension E for the theory, and then says that ϕ is a defeasible consequence if and only if $\phi \in E$. On the "skeptical" or "cautious" strategy, one endorses a conclusion ϕ if and only if ϕ is contained in *every* extension of the theory.

Both the credulous strategy and the skeptical strategy have problems. The problem with the credulous strategy is that the choice of E is arbitrary: With the notion of extension introduced by Reiter, extensions are *orthogonal*; of any two distinct extensions, neither one contains the other. Hence there seems to be no principled way to pick an extension over any other one. This has led a number of researchers to endorse the skeptical strategy as a reasonable approach to the problem of defeasible consequence. However, as shown by Makinson, skeptical consequence, as based on Reiter's notion of extension, fails to be cautiously monotonic. To see this, consider the default theory (W, Δ), where W is empty, and Δ comprises the two defaults

$$\frac{:\theta}{\theta}, \qquad \frac{\theta \vee \eta : \neg\theta}{\neg\theta}.$$

This theory has only one extension, coinciding with the deductive closure of $\{\theta\}$. Hence, if we define $(W, \Delta) \mid\sim \phi$ if and only if ϕ belongs to every extension of (W, Δ), we have $(W, \Delta) \mid\sim \theta$, as well as $(W, \Delta) \mid\sim \theta \vee \eta$ (by the deductive closure of extensions). Now consider the theory with Δ as before, but with W containing the sentence $\theta \vee \eta$. This theory has two extensions: one the same as before, but also another one coinciding with the deductive closure of $\{\neg\theta\}$, and hence not containing θ. It follows that the intersection of the extensions no longer contains θ, so that $(\{\theta \vee \eta\}, \Delta) \not\mid\sim \theta$, against Cautious Monotony.

Given this failure of Cautious Monotony, one would think that the arbitrariness of the credulous alternative might be a small price to pay to preserve one of Gabbay's crucial features identified at the end of Section 1.2. But this hope is short lived, as soon as we notice that the very same theory provides a counterexample for Cut for the credulous strategy.

For suppose we pick the extension of $(\{\theta \vee \eta\}, \Delta)$ that contains $\neg\theta$. Then $(W, \Delta) \mid\sim \theta \vee \eta$ and $(\{\theta \vee \eta\}, \Delta) \mid\sim \neg\theta$, but $(W, \Delta) \not\mid\sim \neg\theta$, against Cut.

It is clear that the issue of how to define a nonmonotonic consequence relation for Default Logic is intertwined with the way that *conflicts* are handled. The problem of course is that neither the skeptical nor the credulous strategy yields an adequate relation of defeasible consequence. In Chap. 3 a notion of general extension for Default Logic is introduced, showing that this notion yields a well-behaved relation of defeasible consequence that satisfies all four requirements of Supraclassicality, Reflexivity, Cut, and Cautious Monotony.

A different set of issues arises in connection with the behavior of Default Logic from the point of view of computation. As we have seen for a given semidecidable set Γ of sentences, the set of all ϕ that are a consequence of Γ in FOL is itself semidecidable. In the case of Default Logic, to formulate the corresponding problem one extends (in the obvious way) the notion of (semi-)decidability given in Section 1.1 to sets of defaults. The problem then is to decide, given a default theory (W, Δ) and a sentence ϕ whether $(W, \Delta) \mid\sim \phi$, where $\mid\sim$ is defined, say, skeptically. Such a problem is not even semidecidable, the essential reason being that, in general, to determine whether a default is triggered by a pair of sets of sentences, one has to perform a consistency check. But the consistency checks are not the only source of complexity in Default Logic. For instance, we could restrict our language to conjunctions of atomic sentences and their negations (making consistency checks feasible). Even so, the problem of determining whether a given default theory has an extension would still be highly intractable (NP-complete, to be precise, as shown by Kautz and Selman, 1991), seemingly because the problem requires checking all possible sequences of firings of defaults. (An accessible introduction to these complexity issues and related notions can be found in Urquhart, 2004.)

In the remainder of this chapter we consider again some of the issues previously introduced in connection with this or that defeasible formalism, but we aim to deal with them from a more abstract point of view.

1.4 SKEPTICAL VERSUS CREDULOUS REASONING

We have already seen that a number of defeasible formalisms allow for a distinction between a credulous and a skeptical approach. This is a distinction that, in general, deals with the issue of how conflicts between potential defeasible conclusions are to be handled. It is therefore worth spending some time looking at the issue of conflicts from a more general point of view.

There are two different kinds of conflicts that can arise within a given nonmonotonic framework: (1) conflicts between defeasible conclusions and "hard facts" and (2) conflicts between one potential defeasible conclusion and another (many formalisms, such as Default Logic, provide some form of defeasible inference rules, and such rules might have conflicting conclusions). When a conflict (of either kind) arises, steps have to be taken to preserve or restore consistency.

All defeasible formalisms handle conflicts of the first kind in the same way: Indeed, it is the very essence of defeasible reasoning that conclusions can be retracted when new facts are learned. But, as we have seen, conflicts of the second kind can be handled in two different ways: One can draw inferences either in a credulous or skeptical fashion (also referred to in the literature as "bold" or, respectively, "cautious"). These two options correspond to widely different ways to construe a given body of defeasible knowledge and yield different results as to what defeasible conclusions are warranted on the basis of such a knowledge base.

The difference between these basic attitudes comes to this. In the presence of potentially conflicting defeasible inferences (and in the absence of further considerations such as specificity – which are in any case typical of a restricted class of formalisms such as inheritance networks), the credulous reasoner always commits to as many defeasible conclusions as possible, subject to a consistency requirement, whereas the skeptical reasoner withholds assent from potentially conflicted defeasible conclusions.

A famous example from the literature, the so-called "Nixon diamond," will help make the distinction clear. Suppose our knowledge base contains (defeasible) information to the effect that a given individual, Nixon, is both a Quaker and a Republican. Quakers, by and large, are pacifists, whereas Republicans, by and large are not. The question is, what defeasible conclusions are warranted on the basis of this body of knowledge, and in particular whether we should infer that Nixon is a pacifist or that he is not pacifist. Figure 1.2 provides a schematic representation of this state of affairs in the form a (defeasible) network.

The credulous reasoner has no reason to prefer either conclusion ("Nixon is a pacifist;" "Nixon is not a pacifist") to the other one, but will definitely commit to one or the other, choosing in some arbitrary fashion. The skeptical reasoner recognizes that this is a conflict not between hard facts and defeasible inferences, but between two different defeasible inferences. Because the two possible inferences in some sense "cancel out," the skeptical reasoner will refrain from drawing either one.

Whereas many of the early formulations of defeasible reasoning have been credulous, skepticism has gradually emerged as a feasible

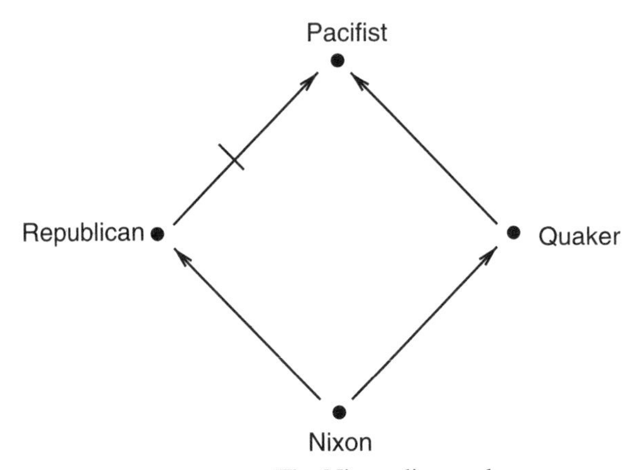

FIGURE 1.2. The Nixon diamond.

alternative, which can, especially for nonmonotonic networks, be better behaved. Arguments have been given in favor of both skeptical and credulous inferences. Some people have argued that credulity seems better to capture a certain class of intuitions, whereas others have objected that although a certain degree of "jumping to conclusions" is by definition built into any nonmonotonic formalism, such jumping to conclusions needs to be regimented, and that skepticism provides precisely the required regimentation. The account of defeasible reasoning developed in this book, as will become increasingly clear, favors a skeptical approach.

1.5 FLOATING CONCLUSIONS

A further issue, related to the skeptical–credulous debate, is the question of whether so-called *floating conclusions* should be allowed. The issue was first raised by Makinson and Schlechta (1991), who claimed that any *direct* approach to skepticism (such as the one in Horty, Thomason, and Touretzky, 1990) was bound to miss certain "floating conclusions." Makinson and Schlechta then argued that the only adequate formalization of skepticism is *indirect* through the intersection of extensions (computational considerations notwithstanding), and a recent rebuttal is due to Horty (2002).

To illustrate the issue, consider, following Horty (2002), the version of the Nixon diamond given in Fig. 1.3. In the figure, the Nixon diamond is supplemented with the information that Republicans tend to be hawks and Quakers tend to be doves, no hawks are doves, and both hawks and

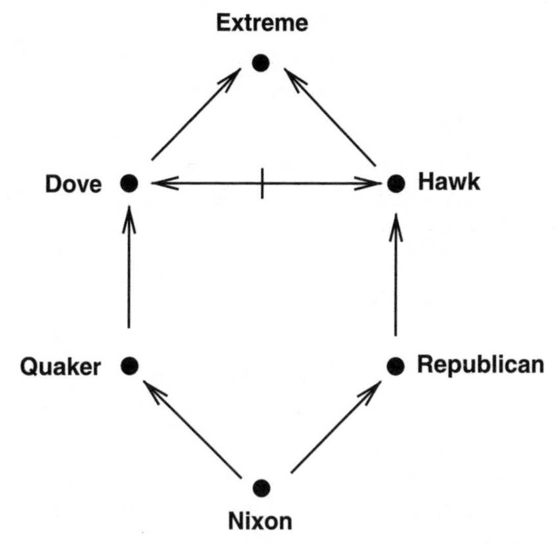

FIGURE 1.3. Floating conclusions in the Nixon diamond.

doves are politically extreme. The question, then, is whether we should infer from this that Nixon is politically extreme.

There is, Horty admits, a *prima facie* argument for a positive answer. Just like the standard Nixon diamond, the network has two extensions, one that endorses the conclusion that Nixon is a hawk and one that endorses the conclusion that he is a dove, and neither one endorsing both. Both extensions, however, endorse the conclusion that Nixon is extreme, so that this conclusion seems to be reached by some sort of "argument by cases," analogous to the following argument pattern in classical sentential logic:

$$\frac{\begin{array}{c} A \vee B \\ A \rightarrow C \\ B \rightarrow C \end{array}}{C}\,.$$

The conclusion then seems unassailable even if one adheres to a skeptical approach. So far, the intuition. The issue is how such an intuition is to be implemented by a mechanism that deals with the network formalism directly. A first option is to use a *direct* approach that identifies a privileged extension of the network; if such an extension implements skepticism, neither path, "Nixon → hawk → extreme" or "Nixon → dove → extreme," will be in it, so that the conclusion that Nixon is extreme is lost.

An alternative approach to skepticism follows the *indirect* route, first constructing all the credulous extensions of the network, and then taking their intersection. There is a possible ambiguity here, as it is not clear whether one first takes the intersection of the extensions (construed as sets of *paths*), and *then* moves to the set of conclusions they support, or else one considers, for each extension, the set of conclusions it supports, and *then* takes the intersection of the resulting sets.

The two operations, clearly, do not commute. Only the latter strategy yields the conclusion that Nixon is extreme. Each of the two credulous extensions of the net contains a path from "Nixon" to "Extreme," but there is no common path connecting the two nodes. So according to the former strategy, if we take the intersection of the two extensions of the net, again the conclusion that Nixon is extreme is lost. The statement that "Nixon is extreme" is therefore a floating conclusion. The conclusion that Makinson and Schlechta endorse (along with other authors) is that the proper implementation of skepticism consists in the latter version of the indirect route.

Their argument, though, relies on the implicit assumption that the floating conclusion should indeed be endorsed. But an argument to the contrary can be found in Horty (2002). Horty (2002, p. 62) asks us to consider the following story (whose structure is depicted in Fig. 1.4):

> Suppose that my parents have a net worth of one million dollars, but that they have divided their assets in order to avoid the United States inheritance tax, so that each parent currently possess half a million dollars apiece. And suppose that, because of their simultaneous exposure to a fatal disease, it is now settled that both of my parents will die within a month. This is a fact: medical science is certain.
>
> Imagine also, however, that there is some expensive item – a yacht, say – whose purchase I believe would help to soften the blow of my impending loss. Although the yacht I want is currently available, the price is good enough that it is sure to be sold by the end of the month. I can now reserve the yacht for myself by putting down a large deposit, with the balance due in six weeks. But there is no way I can afford to pay the balance unless I happen to inherit at least half a million dollars from my parents within that period, and if I fail the pay the balance on time, I will lose my large deposit. Setting aside any doubts concerning the real depth of my grief, let us suppose that my utilities determine the following conditional preferences: if I believe I will inherit half a million dollars from my parents within six weeks, it is very much in my benefit to place a deposit on the yacht; if I do not believe this, it is very much in my benefit not to place a deposit.
>
> Now suppose I have a brother and a sister, both of whom are extraordinarily reliable as sources of information. Neither has ever been known to be mistaken, to deceive, or even to misspeak – although of course, like nearly any source of information, they must be regarded as defeasible. My brother and sister have

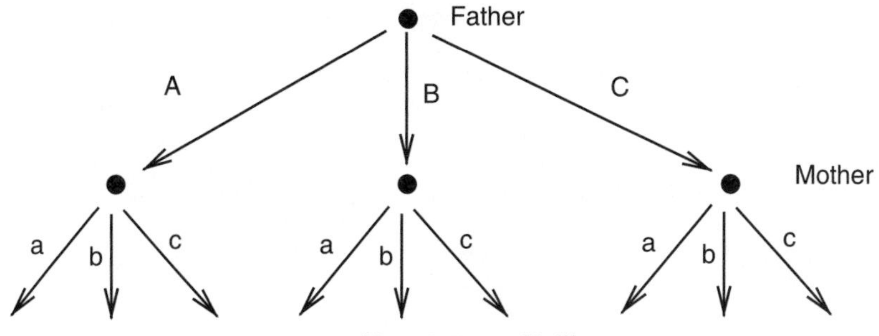

FIGURE 1.4. Horty's "moral" dilemma.

both talked with our parents about their wills, and feel that they understand the situation. I have written to each of them describing my delicate predicament regarding the yacht, and receive letters back. My brother writes: "Father is going to leave his money to me, but Mother will leave her money to you, so you're in good shape." My sister writes: "Mother is going to leave her money to me, but Father will leave his money to you, so you're in good shape." No further information is now available: the wills are sealed, my brother and sister are trekking together through the Andes, and our parents, sadly, have slipped into a coma.

The question, then, is whether I should go ahead and put down the deposit. The situation is depicted, in decision-theoretic terms, in Fig. 1.4. Here, father's three possible courses of action, i.e., leaving his money to me, to my brother, or to my sister, are represented as A, B, and C, respectively. Similarly, mother's three possible courses of action are represented as a, b, and c. Accordingly, my brother asserts "B and a" and my sister asserts "A and c."

Intuitions, in this case, are markedly different from the extreme Nixon case. Whereas there the floating conclusion seemed unescapable, here there is a strong intuition that it would be foolhardy of me to rush to put down the deposit. And yet, as Fig. 1.5 makes clear, the statement that I will inherit a large sum of money is a floating conclusion, and in fact the argument has the very same structure as in the Nixon case.

The conclusion that Horty reaches is that floating conclusions are not always desirable, and that one should not want a formalism that automatically forces one to accept them. It is indeed possible to go one step further. Once the intuition that floating conclusions are acceptable has been undermined, one can see that also the most prominent example in its favor (the modified Nixon diamond of Fig. 1.3) is subject to criticism. Suppose that indeed we espouse a genuinely skeptical standpoint. On this view, what reasons are really there to conclude that Nixon is politically extreme? There are, as we know, two possible arguments that can

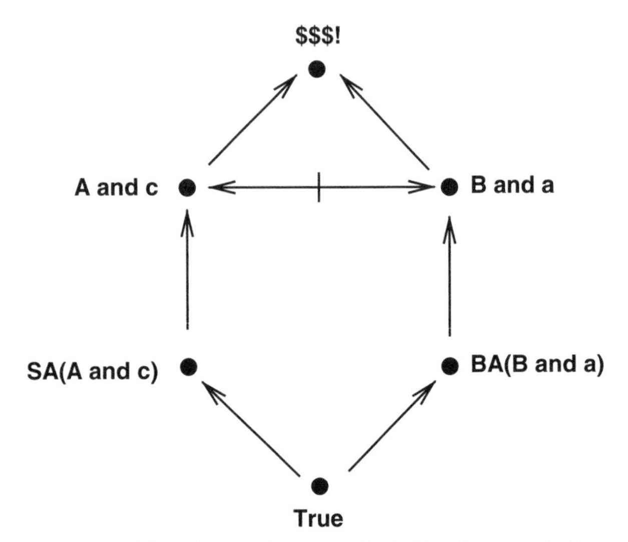

FIGURE 1.5. Horty's counterexample to floating conclusions.

be given in support of such a conclusion. The first argument proceeds by way of the fact that Nixon is a Republican and hence a hawk; but the argument is undermined by the possibility that Nixon, as a Quaker, might be a dove after all. The second argument proceeds by way of the fact that Nixon is a Quaker and hence a dove; but the argument is undermined by the possibility that Nixon, as a Republican, might be a hawk after all. Do we still think that we have any reason to conclude that Nixon is politically extreme? It seems clear that we have no such reason, and that therefore the conclusion should, from a skeptical point of view, be rejected.

1.6 CONFLICTS AND MODULARITY

Another conceptual issue that arises in nonmonotonic formalisms concerns the *nature* of conflict. Whereas quite a bit of literature is devoted to the issue of how conflicts should be handled (the skeptical–credulous debate falls under this rubric), the issue of what *constitutes* conflict is often neglected.

In fact, this is an important issue that resurfaces every now and then, except that it is rarely recognized for what it is, and the questions it raises are often handled as though they concerned the *manner* in which conflicts are to be handled.

An example of this can be found, for example, in Brewka and Gottlob (1997). In that paper, Brewka and Gottlob propose a variant of Default

Logic semantics that is based on the so-called "well-founded semantics" (WFS) for logic programs. In WFS (proposed by van Gelder, Ross, and Schlipf, 1991, and further developed by Baral and Subrahmanian, 1991; Przymusińska and Przymusiński, 1994) one exploits the fact that the double iterate of Reiter's original operator $\mathfrak{S}$ (Section 1.3) is indeed monotonic, and hence has a least fixpoint. In this section we are mostly concerned with broad conceptual issues, and hence we do not deal with the technical details of well-founded semantics. But some of its features are of considerable foundational interest.

Brewka and Gottlob identify four advantages that well-founded semantics enjoys over Reiter's original approach (1997, p. 2):

1. Existence of a least fixpoint – every theory gives rise to a reasonable consequence relation.
2. The least fixpoint can be approximated from below.
3. Cumulativity: Adding a skeptical conclusion to the premises does not change the set of skeptical conclusions.
4. In at least some important cases, the set of skeptical conclusions can be efficiently computed.

This last item is obviously very important, but it is not dealt with here. Rather, the focus is on the first three (which, incidentally, are all satisfied by the approach to general extensions introduced in Chap. 3). But WFS suffers from several drawbacks, one of which is the fact that at times *no conclusions* can be inferred from a default theory (this is what prompts Brewka and Gottlob to propose their own variant of WFS, with a characteristically "syntactic" or "proof-theoretic" flavor – the details are not given here).

But another drawback of WFS, for Brewka and Gottlob, is the fact that WFS is too *conservative*. For instance, consider the default theory comprising the following defaults:

$$\frac{:B}{B}, \qquad \frac{:A}{A}, \qquad \frac{:\neg A}{\neg A}.$$

Although the last two defaults quite clearly conflict each other and hence cannot be triggered, nothing should prevent triggering the first default. But the least fixpoint extension of WFS is empty, and no default is triggered. From this example, Brewka and Gottlob proceed to change the monotone operator slightly, using the notion of a default proof. In particular they employ, in an essential way, sets of sentences that are not deductively closed. This allows them to "keep inconsistencies local" and therefore to obtain the desired results in the preceding theory.

Indeed, the same point about local conflicts can be made with respect to the somewhat simpler default theory comprising the two defaults

$$\frac{:B}{B}, \qquad \frac{:A}{\neg A}.$$

Here again we have a "self-defeating" default (the second one), which should not, however, prevent the triggering of the first. In this case as well, we can make the case that conflicts should be kept local.

There is big difference, however, between the first and second example of nonlocality. In the second case, we have that the conclusion of the second default defeats the justification of the same default. In the first case, instead, besides the conflict between the last two defaults (each of which defeats the justification of the other) we also have that the two sentences A and $\neg A$ *together* imply $\neg B$ and therefore defeat the justification B of the first default, preventing it from being triggered. This is quite a different phenomenon from $\neg A$ defeating A, and one that needs to be addressed separately.

To address this issue, Brewka and Gottlob resort to *nondeductively closed* sets of sentences, a fact that gives their approach its characteristic syntactic flavor. Instead, one would like to see a more *modular* approach, in the sense of taking some kind of consequence relation between sentences (the relation $\models$ of classical logic, say) as basic, and build a relation $\mid\!\sim$ of defeasible consequence on top of it.

This leads to a clear division of labor. The basic consequence relation (in our case, $\models$) is in charge of defining what counts as conflict, whereas the defeasible relation $\mid\!\sim$ is in charge of defining how conflicts are to be handled. The two tasks are conceptually distinct, and so they should remain. Moreover, such an approach is modular because one can in principle change the underlying basic relation and use a different notion of conflict without having to change the intuitions behind the way conflicts are handled. For instance, one could imagine using a consequence relation that embodies certain *relevant* considerations (such as the four-valued logic of Belnap, 1977). On the basis of such a consequence relation, then, the sentences A and $\neg A$ would no longer defeat B, but they would still defeat each other.

In other words, it can be reasonably claimed that the problem with Brewka and Gottlob's first example is not (or not only) a problem with the way that extensions are built up from $\models$, but also a problem with $\models$ itself and the blatant failures of relevance it engenders. Indeed, as was already noted in Section 1.2, relevance seems to play a greater role in defeasible reasoning than hitherto acknowledged. Indeed, what the approach of Brewka and Gottlob amounts to (and as they explicitly acknowledge) is a localization of inconsistencies, a task that has kept the relevant logicians busy for over 40 years (see Anderson, Belnap, and Dunn, 1975–1992, for an extensive treatment). But Brewka and Gottlob do so by mixing the two conceptually different tasks of identifying conflicts and handling them: Localizing conflicts is a task that is largely independent of that of building a consequence relation for Default Logic. Any approach to consequence for Default Logic needs to pursue a modular strategy, by first identifying a notion of "conflict," and then devising the mechanisms necessary to handle it.

In fact, there is an approach to relevance that can be extracted from the approach of Brewka and Gottlob: Where S is a (possibly inconsistent) set of sentences, define "S entails ϕ" iff ϕ belongs to the deductive closure of a maximally consistent subset of S. What Brewka and Gottlob are doing is taking WFS and replacing the relation $\models$ of classical logic with this relation, "entails," throughout, reaping some of the advantages of the added relevance. But a case can be made that, if one really wanted relevance, it might be better to take a more modular approach.

Perhaps one more consideration is suggested by the preceding discussion. For a long time, Anderson, Belnap, and Dunn and their students have been pointing out the outrageous failures of relevance found in classical logic, but their words have found many deaf ears. Possibly because of the complications implicit in a relevant approach, many people are quite happy living with such failures. But such failures become problematic in a defeasible setting: Although relevance failures are more or less readily accepted in classical logic, all of a sudden they leap to the foreground in a defeasible setting. It might well be because failures of relevance become more noticeable in a new setting such as that of defeasible reasoning; or it might be that there really is something about defeasible reasoning that cries out for relevance more than in classical logic. Be that as it may, a modular approach allows us clearly to distinguish what is due to the underlying notion of conflict from what is due to the strategy we have chosen to handle those conflicts.

1.7 ASSESSMENT

There are three major issues connected with the development of logical frameworks that can adequately represent defeasible reasoning: (1) material adequacy, (2) formal properties, and (3) complexity. Material adequacy concerns the question of how broad a range of examples is captured by the framework and the extent to which the framework can do justice to our intutions on the subject (at least the most entrenched ones). The question of formal properties has to do with the degree to which the framework allows for a relation of logical consequence that satisfies the conditions of Supraclassicality, Reflexivity, Cut, and Cautious Monotony. The third set of issues has to do with computational complexity of the most basic questions concerning the framework.

There is a potential tension between (1) and (2): The desire to capture a broad range of intuitions can lead to ad hoc solutions that can sometimes undermine the desirable formal properties of the framework. In general, the development of nonmonotonic logics and related formalisms has been driven, since its inception, by consideration (1) and has relied on a rich and well-chosen array of examples. Of course, there is some question as to whether any single framework can aspire to be universal in this respect.

More recently, researchers have started paying attention to consideration (2), looking at the extent to which nonmonotonic logics have generated well-behaved relations of logical consequence. As Makinson (1994) points out, practitioners of the field have encountered mixed success. In particular, one abstract property, Cautious Monotony, appears at the same time to be crucial and elusive for many of the frameworks found in the literature. This is a fact that is perhaps to be traced back, at least in part, to the previously mentioned tension between the requirement of material adequacy and the need to generate a well-behaved consequence relation.

The complexity issue appears to be the most difficult among the ones that have been singled out. Nonmonotonic logics appear to be stubbornly intractable with respect to the corresponding problem for classical logic. This is clear in the case of Default Logic, given the ubiquitous consistency checks. But beside consistency checks, there are other, often overlooked, sources of complexity that are purely combinatorial. Other forms of nonmonotonic reasoning, besides Default Logic, are far from immune from these combinatorial roots of intractability. Although some important work has been done in trying to make various nonmonotonic formalism more tractable, this is perhaps the problem in which progress has been slowest in coming.

2

Defeasible Inheritance over Cyclic Networks

2.1 BACKGROUND AND MOTIVATION

Defeasible inheritance networks were originally developed to gain a sound mathematical understanding of the way inheritance systems store, gain access to, and manipulate taxonomic information with exceptions (a survey can be found in Thomason, 1992). An inheritance network can be identified with a collection of *signed links* (positive or negative) over a set of nodes. Such links are of the form $n_1 \rightarrow n_2$ or $n_1 \not\rightarrow n_2$, respectively, where n_1 and n_2 are nodes in the net. Such nodes are labeled by lexical items referring to categories of individuals. It is convenient to identify nodes with their labels. When the network is to be interpreted defeasibly, a link $n_1 \rightarrow n_2$ represents the fact that objects of category n_1 tend to be of category n_2, whereas a link of the form $n_1 \not\rightarrow n_2$ represents the fact that objects of category n_1 tend *not* to be of category n_2. A *path* over a net Γ is a sequence of links from Γ, at most the last one of which is allowed to be negative. So both $n_1 \rightarrow n_2 \rightarrow n_3$ and $n_1 \rightarrow n_2 \not\rightarrow n_3$ are paths, whereas $n_1 \not\rightarrow n_2 \rightarrow n_3$ is not. A path is positive or negative according to whether its last link is positive or negative.

Theories of defeasible inheritance found in the literature follow either the *direct* approach (defining a notion of *consequence* for inheritance networks in terms of the net itself), or the *indirect* approach (which assigns meaning to inheritance networks by embedding them in a language already equipped with well-understood semantics). For instance, an indirect approach was pursued, in the case of *strict* inheritance networks, by Hayes (1979) (by means of an embedding into FOL), and in the case of defeasible inheritance networks by Etherington and Reiter (1983) (using an embedding into default logic). However, the *direct* approach first introduced by Touretzky (1986) has now become standard, and that is the point of view adopted here.

Inheritance theories are based on the three fundamental notions of *constructibility*, *conflict*, and *preemption*. Roughly speaking, we say that a path is *constructible* relative to a net Γ if it can be obtained by chaining links from Γ in a forward fashion. A path *conflicts* another path containing at least two links if the first has the same endpoints but opposite sign as the the second. So $n_1 \rightarrow n \rightarrow n_2$ is conflicted by $n_1 \rightarrow n' \not\rightarrow n_2$, and conversely. But perhaps the most important idea in defeasible networks, and the one that is meant to embody our intuitions concerning inheritance, is that of *preemption*. Preemption gives us a way to resolve conflicts between paths, based on the intuition that more specific information should override more generic information.

There are two ways to define preemption: *On-path* preemption, originally proposed by Touretzky (1986) and Boutilier (1989), and *off-path* preemption of Sandewall (1986), Horty et al. (1990), and Stein (1992), which has come to be prominent in the literature and is used in this chapter. Consider for instance one of the standard examples represented in the net of Fig. 2.1. Here, although we are told that Tweety is a penguin, penguins are birds, and birds fly, the conclusion we naturally draw is that Tweety does not fly. The conclusion that Tweety flies is preempted by information to the effect that penguins don't fly. Because penguins are a kind of bird, information as to whether penguins fly is more specific than information about whether birds fly, and thus overrides it. We conclude that Tweety does not fly. The notion of preemption captures this formally, using only topological properties of the network itself.

Once the notions of constructibility, conflict, and especially preemption have been defined, we can proceed with the definition of the *extensions* of a net Γ. Intuitively, an extension is a conflict-free set of paths that are supported by the net. There are essentially two ways to define extensions:

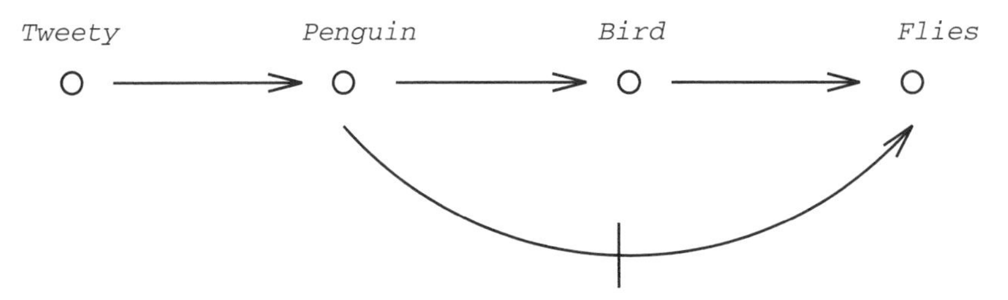

FIGURE 2.1. The standard example of preemption.

credulous (see Touretzky, 1986) and *skeptical* (see Horty et al., 1990; a third approach based on *flexible* extensions, is described in Horty (1994), but is ignored here). Although skeptical extensions might be preferable on conceptual and computational grounds (as argued, among others, in Horty et al., 1990 and Simonet and Ducournau, 1994), the former is somewhat simpler. The reader is again referred to the excellent survey by Horty (1994) for details on these two kinds of extensions. As a first approximation we say that an extension for a net Γ is credulous if it endorses a *maximal* conflict-free set of paths over Γ in which no path is preempted; whereas skeptical extensions only endorse, among the "unpreempted" paths, those that are not in turn conflicted by other "unpreempted" paths.

As with other "consistency-based" approaches to defeasible reasoning (such as Default Logic), extensions need not be unique. However, it is well known that, if the underlying net Γ contains no cycles, then extensions always exist. This can be seen as follows. Given a path σ over Γ, define the *degree* of σ to be the length of the longest sequence of links (irrespective of their signs) from Γ having the same endpoints as σ. It is clear that this notion of degree makes sense only if Γ contains no cycles. Then, in the case of acyclic nets, it is possible to show that extensions exist by means of an iterative process in which paths are considered in ascending order of their degrees (see Horty, 1994, for a unified treatment of inheritance on acyclic nets).

Things are different in the case of networks with cycles. Such nets arise naturally in many situations, for instance whenever there are two mutually overlapping categories (see Fig. 2.2 for an example). In such nets, the presence of cycles is a cause for the *global* character of the notion of specificity. As long as the net is acyclic, if a node n_1 is more specific than node n_2, then this character is preserved no matter how the network is extended. In this sense, in acyclic contexts, specificity is a *local* property of nodes. But once cycles are allowed, however, a node n_1 might be more specific than a node n_2 relative to a certain net Γ, but not relative to a net Γ' extending Γ (because Γ' might introduce a path back from n_2 to n_1). In this sense, specificity is a global property of nodes in cyclic nets.

This appears to be connected with the fact that cyclic networks need not have extensions in the standard sense, as it was discovered by Horty (1994). Consider the net of Fig. 2.3. According to the usual approaches to inheritance, this net cannot have any extensions. Suppose for contradiction that Φ is an extension for the net. Then, clearly, either the path $a \to b \to c$ is in Φ or it isn't. If it is, then Φ must contain also the path

$$\sigma = a \to b \to c \to d \to e \to b$$

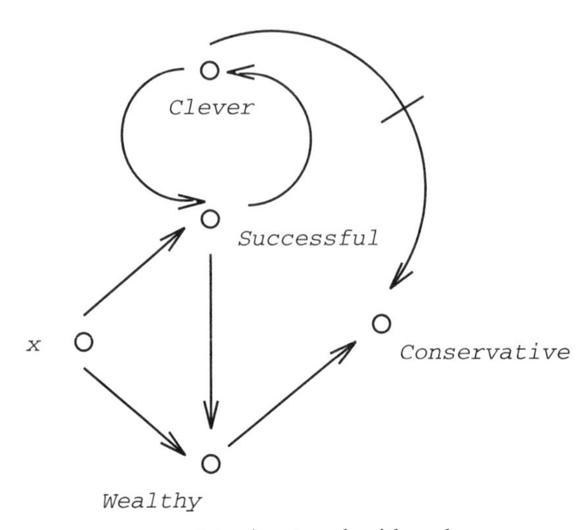

FIGURE 2.2. A network with cycles.

(because nodes d, e, b have no incoming negative links, neither conflict nor preemption can block the construction of σ); but then $a \to b \to c$ would be preempted in Φ, as Φ contains a node e that is more specific than b (because the path σ is in Φ) and a direct link telling us that es are not cs. This is impossible if Φ is an extension. If $a \to b \to c$ is not in Φ then σ can't be in Φ either, so that $a \to b \to c$ would not be preempted in Φ and so Φ cannot be an extension (because it would fail to contain a path that is constructible but neither conflicted nor preempted in Φ). The reason for this state of affairs seems to be that the path $a \to b \to c \to d \to e \to b$ preempts one of its initial segments.

As noted (Section 1.3), the situation with defeasible networks is not too dissimilar from the semantic paradoxes such as the Liar. If we construe

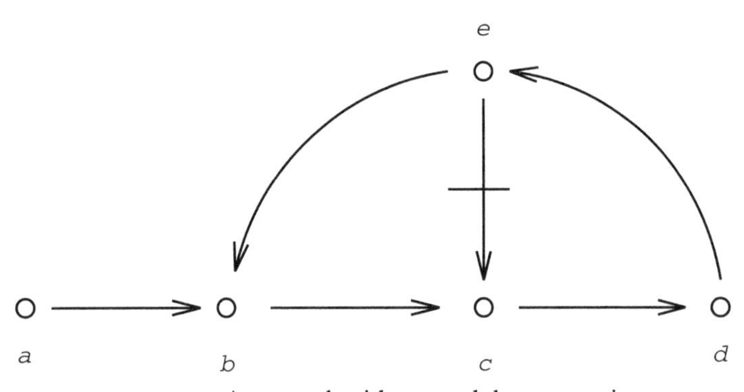

FIGURE 2.3. A network with no credulous extension.

paths as arguments, then in the context of the net of Fig. 2.3 the path $a \to b \to c \to d \to e \to b$ says of itself that it is not tenable. It is interesting to notice, however, that no explicit self-reference is anywhere in sight. We know from the discussion of Fig. 2.3 that we have to give up the Touretzky–Horty notion of extension if we are to deal with cyclic networks.

In what follows we are going to provide a solution to this problem by defining a notion of extension according to which all nets have extensions. Such a solution is indeed inspired by the Kripke (1975) construction of a semantics for a language containing its own truth predicate. In that construction, given a three-valued truth schema such as the "strong Kleene" (of Kleene, 1952) – in which $\neg\varphi$ is true, false, or indeterminate according to whether φ is false, true, or indeterminate, respectively – a sentence φ no longer needs explicitly to be counted as true in order to prevent $\neg\varphi$ from being counted as true; and similarly, it is only when explicitly counted as false that a sentence φ can no longer prevent $\neg\varphi$ from being counted as true.

The intuition behind the present approach to cyclic nets applies this idea to the definition of the concept of extension. An extension, according to present proposal, is a pair of sets of paths (the paths that are explictly constructed and the ones that are explicitly ruled out) simultaneously satisfying a pair of fixpoint equations. It is then no longer necessary for a path to be explicitly constructed for it to *preempt* other paths. On the other hand, once a path has been explicitly ruled out as preempted or conflicted, it can no longer preempt other paths.

Similar approaches have been pursued by Makinson and Schlechta, but for other reasons. In particular, Makinson and Schlechta (1991, p. 206) consider possible ways to allow *zombie paths* (paths that are neither explicitly ruled out nor explicitly allowed) to do their work (see also Schlechta, 1993). A solution they consider (but only in order later to reject it) is to allow "intermediate" degrees of path constructions, between "explicitly allowed" and "explicitly ruled out." They show that as long as we pursue a localized, forward chaining approach, no increase in the (finite) number of intermediate values will allow zombie paths. We will come back to this issue in Subsection 2.5.2, but for now we notice, for future reference, that the nonlocal character of the approach given in this chapter circumvents this line of objection.

The similarities between the present approach to cyclic nets and the three-valued solution to the Liar Paradox are exploited in order to single out, among the extensions introduced in this chapter, those that are referred to as *classical* in that they are the analog of two-valued semantics for a formal language. In a classical extension any path not explicitly constructed is explicitly ruled out, so that there are no paths that fall *in between*. Every extension in the sense of Touretzky–Horty is a classical extension in our sense. In particular, the Touretzky–Horty notion of extension is subsumed under the notion of extension presented here.

It is well known that every *acyclic* net has a classical extension. But these are not the only nets that have classical extensions. As we will see in Section 2.4, there are cyclic nets that exhibit a behavior similar to that of the Truth-Teller. Such nets are the $(2n)$ loops of Definition 2.4.6, and Theorem 2.4.3 establishes – somewhat surprisingly – the existence of classical extensions for them. It is important to notice that the length (even or odd) of the cycles seems to play a crucial role in determining whether a net has a classical extension: Although every net has an extension, whether it contains odd- or even-length cycles, such an extension will not, in general, be classical.

Although the present approach draws its inspiration from the analogy with solutions of the Liar Paradox given in philosophical logic, it is connected to other approaches in the literature. First, there are direct approaches dealing with cycles in *strict* inheritance networks (see Thomason, Horty, and Touretzky, 1987). When the network is construed strictly, cycles pose no particular problem (this is not to say that the semantics for such a network is trivial: on the contrary, it is unexpectedly complex, but the complexity appears to be independent of whether the net contains cycles).

A different approach to cycles, although not in inheritance networks but in *terminological systems*, is given by Nebel (1990). Nebel considers the case of a sequence of definitions, in which some lexical item is defined either directly in terms of itself or indirectly in terms of other items that in turn are defined in terms of it. Again, Nebel's construction takes place in a strict setting and proceeds by finding the least fixpoints of certain monotonic operators. Such fixpoints are then interpreted as providing the classical extension of the lexical items being defined. No paradoxical phenomena force the adoption of a "three-valued" approach. A nonmonotonic extension of terminological systems is considered in Baader and Hollunder (1994): They propose a merge of terminological systems and a particular version of Reiter's Default Logic, but they do not address the particular problems deriving from cyclic representation formalisms. Finally, an indirect approach to defeasible networks with cycles by means of a translation into logic programs can be found in You, Wang, and Yuan (1999).

On the other hand, there is an interesting connection between the present proposal and work in Default Logic by Papadimitriou and Sideri (1994). Building on previous work by Etherington (1987), they show how to associate with any default theory a particular graph, representing the logical dependencies among the defaults comprising the theory, and then they establish that the theory must have at least one extension (in the sense of Default Logic; see, for instance, Reiter, 1980), provided the associated graph contains no odd-length cycles.

It seems that the same sort of phenomenon is at work in cyclic inheritance networks as in default theories: In both cases we have a sequence of what could be regarded as *inference rules*, the firing of each one of which prevents triggering the next, and the firing of the last one of which prevents triggering the first (defaults are clearly sorts of inference rules, and paths through a net can also be construed as inference rules; for instance, this is the point of view of Horty, 1994). If the cycle has an even length, it is possible consistently to partition the sequence in two alternating subsequences, containing the rules that are triggered and the rules that are preempted. If the cycle has an odd length, no such partition is possible.

Moreover, the notion of extension employed in default logic is intrinsically two-valued in the sense that it contains the consequences of a maximal set of defaults whose justifications are consistent with the extension itself. In other words, the triggering of a default can be prevented only if its justification is explicitly refuted. In virtue of this maximality Papadimitriou and Sideri fail to consider a possibility that is available to the present approach to inheritance networks, namely, that the subsequences of the rules mentioned in the previous might fail to be exhaustive. This would indeed give rise to a three-valued notion of extension for Default Logic, analogous to the one put forward here for inheritance networks, and according to which every default theory has an extension. Such a notion of extension is developed in Chap. 3.

2.2 GRAPH-THEORETICAL PRELIMINARIES

A finite inheritance network is a two-sorted directed graph, i.e., a pair consisting of a finite set of nodes $\{n_1, \ldots, n_p\}$, along with a finite set of *signed* links between nodes of the form $n_i \rightarrow n_j$ or $n_i \not\rightarrow n_j$. Because Γ contains finitely many links, there are also finitely many links between any two nodes: We can then assume with no loss in generality that given two nodes there is at most one positive and at most one negative link between the first and the second. A sequence of n nodes is a *trail* if either $n = 0$ or $n \geq 2$, and moreover each node in the sequence is connected by a link to the next one.

The following notational conventions are used. Lowercase letters of the Greek alphabet, $\rho, \sigma, \tau, \ldots$, denote trails, and uppercase letters of the Greek alphabet, $\Phi, \Psi, \ldots$, represent sets of trails. We use λ to refer to the empty trail. Negative links are represented by a bar over the end node (so for instance $n_i \not\to n_j$ is represented by $n_i\overline{n_j}$); we will abuse the language and use $\overline{\overline{n_i}}$ to denote n_i. Let $\rho = x_1 \ldots x_n a$ and $\sigma = a y_1 \ldots y_m$ be trails. Then the juxtaposition $\rho\sigma$ represents the sequence $x_1 \ldots x_n a y_1 \ldots y_m$. Moreover, if y is the end node of ρ, the notation ρx is shorthand for $\rho\langle y, x\rangle$, and similarly for $x\rho$.

Let $\alpha = x_1 \ldots x_n$. A node x occurs in a trail α if $x = x_j$ for $1 \leq j \leq n$. Given α as previously and a trail β we say that β is a prefix or initial segment (not necessarily proper) of α, written as $\beta \sqsubset \alpha$, if and only if $\beta = x_1 \ldots x_m$ for some $m \leq n$. We also say that α is a subtrail of β if $\beta = \gamma\alpha\delta$ for some (possibly empty) trails γ and δ.

Definition 2.2.1 fixes a technical meaning for the word "path," which is adopted throughout the chapter. In particular, the definition depends on some antecedently fixed net Γ, so the only paths are the ones obtained by chaining links in Γ.

DEFINITION 2.2.1 *A trail is a* path *if and only if it contains at most one negative link, and such a link occurs as the last link in the trail. A path is* positive *or* negative *according to whether its last link is positive or negative. By Γ^* we refer the set of all (positive as well as negative) paths over Γ.*

From the definition it follows that we can have paths of any nonnegative number of nodes except one. If α and β are paths and α is a subsequence of β, we say that α is a *subpath* of β.

DEFINITION 2.2.2 *The* length *of a path σ, denoted by $\ell(\sigma)$, is the number of* links *in σ. For example, if $\sigma = x_0 \ldots x_n$, then the length of σ is n. We also set $\ell(\lambda) = 0$.*

DEFINITION 2.2.3 *If σ is a nonempty path, then σ^{i} and σ^{e} denote its initial and end node, respectively.*

DEFINITION 2.2.4 *A path $\sigma \in \Gamma^*$ is* simple *if every node occurs at most once in it. Let S_Γ refer to the set of simple paths over Γ.*

DEFINITION 2.2.5 *A path $\sigma \in \Gamma^*$ is a* cycle *if it is positive and has the form $x\rho x$, and $x\rho$ is simple. We refer to the set of cycles over Γ as C_Γ. A set of* *paths* Φ *is* cyclic *if some path in* Φ *contains a subpath that is a cycle, and* Φ *is* acyclic *otherwise; a net* Γ *is* cyclic *if and only if* Γ^* *is cyclic.*

If Γ^* contains cycles, then Γ^* will be infinite, independently of whether the underlying network Γ itself is finite. (Observe that negative trails of the form $x\alpha\overline{x}$ do not give rise to an infinity of paths in Γ^*, as there is no chaining off of a negative link.) Given the infinity of Γ^* we are interested in defining a *finite* subset $\Gamma^\#$ of Γ^* with the property that $\Gamma^\#$ will contain a path from a node x to a node y if and only if Γ^* contains a path of the same sign from x to y. This will allow us to reduce questions about Γ^* to questions about $\Gamma^\#$. The basic idea in the construction of $\Gamma^\#$ is to take all the paths with no repetitions from Γ^* and "splice in" cycles in such a way as to "go around" each cycle at most once. Although this idea is subsequently made precise, it should be noted that this is by no means the only possible choice. In general, any finite subset of Γ^* that is closed under initial segments could be used in place of $\Gamma^\#$.

LEMMA 2.2.1 *S_Γ and C_Γ are both finite.*

Lemma 2.2.1 follows immediately from the finiteness of Γ. The following is a definition of "splicing" – the process that splices a cycle into a simple path that intersects the cycle (see Fig. 2.4).

DEFINITION 2.2.6 *Let ρ be a cycle with $\rho^{\mathrm{i}} = \rho^{\mathrm{e}} = x$, and let σ and τ be paths; we say that σ is* obtained from τ by splicing ρ, *written $\tau \leq_\rho \sigma$, if and only if τ is of the form $\alpha x \beta$, and σ is $\alpha\rho\beta$.*

This notion of splicing can be generalized to capture the process by which a path σ can be turned into a path τ by successive splicings of

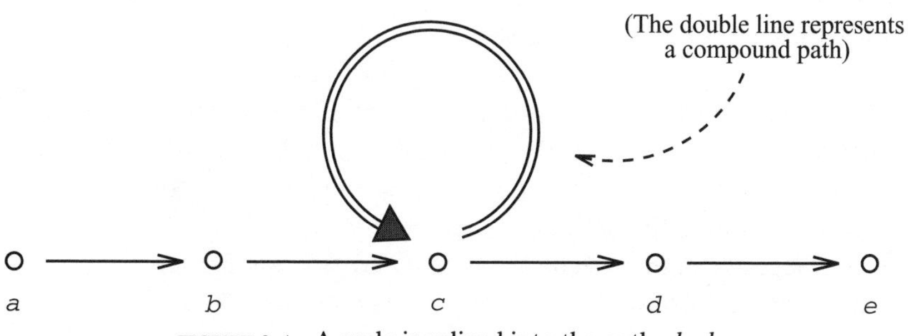

FIGURE 2.4. A cycle is spliced into the path *abcde*.

cycles drawn from C_Γ, but in such a way that each cycle is used at most once.

DEFINITION 2.2.7 *Let $\mathcal{T}$ be the set of all sequences of paths of the form $\langle \lambda, \alpha_1, \ldots, \alpha_{k+1} \rangle$ such that* (1) $\alpha_1 \in S_\Gamma$ *and* (2) *there exist pairwise distinct paths $\rho_1 \ldots \rho_k \in C_\Gamma$ such that $\alpha_i \leq_{\rho_i} \alpha_{i+1}$, for all $i \leq k$.*

$\mathcal{T}$, together with the initial segment relation on sequences of paths, constitutes a tree.

THEOREM 2.2.1 *$\mathcal{T}$ is finite.*

(A precise bound is given in Section 2.6.)

DEFINITION 2.2.8 *Let $\Gamma^\#$ be the set of all paths α occurring in a sequence in $\mathcal{T}$.*

It follows immediately from Definition 2.2.8 that $\Gamma^\#$ is finite.

2.3 CONSTRUCTING EXTENSIONS

Because, as mentioned in Section 2.1, extensions in the creduluous sense of Touretzky (1986) do not necessarily exist for cyclic graphs, we define a new notion of extension that agrees with the definition of credulous extension for acyclic nets and guarantees that cyclic nets also have extensions. In what follows, recall that any nonempty path σ that is not a link can be written in the form τx or $\tau \overline{x}$, where $\tau^e \to x$ or (respectively) $\tau^e \not\to x$ is a link.

DEFINITION 2.3.1 *Let Φ be a set of paths. A path σ is* constructible *in Φ, relative to Γ, if and only if σ is the empty path, or σ is one of the links in Γ, or σ has the form τx or $\tau \overline{x}$, where $\tau \in \Phi$ and the link $\tau^e \to x$ or, respectively, $\tau^e \not\to y$ is in Γ.*

Although links are never conflicted, they can in turn conflict longer, compound paths. This is justified by the intuition that a network Γ should always support at least those statements corresponding to the links. This insight is captured by the following definition, according to which a positive or negative path is conflicted in Φ if and only if it has length ≥ 2 and Φ contains a path with the same endpoints but opposite sign.

DEFINITION 2.3.2 *Let Φ be a set of paths; we say that a path σ is* conflicted *in Φ if and only if $\ell(\sigma) \geq 2$, and*

$$\exists \tau \in \Phi(\tau^{\mathrm{i}} = \sigma^{\mathrm{i}} \,\&\, \tau^{\mathrm{e}} = \overline{\sigma^{\mathrm{e}}}).$$

Observe that given our notational convention that $\overline{\overline{x}} = x$, Definition 2.3.2 captures the notion of conflict for both positive and negative paths simultaneously.

DEFINITION 2.3.3 *A positive path σx (of length ≥ 2) is* preempted *in Φ (relative to Γ) if and only if $\sigma \in \Phi$ and there is a node v such that the link $v \not\rightarrow x$ is in Γ, and either $v = \sigma^{\mathrm{i}}$, or $\sigma^{\mathrm{i}}\tau_1 v \tau_2 \sigma^{\mathrm{e}} \in \Phi$, for some paths τ_1 and τ_2.*

Similarly, a negative path $\sigma\overline{x}$ is preempted *in Φ (relative to Γ) if and only if $\sigma \in \Phi$, and there is a node v such that the link $v \rightarrow x$ is in Γ, and either $v = \sigma^{\mathrm{i}}$, or $\sigma^{\mathrm{i}}\tau_1 v \tau_2 \sigma^{\mathrm{e}} \in \Phi$, for some paths τ_1 and τ_2.*

A diagram representing this notion of preemption can be found in Fig. 2.5 (where the thick lines represent possibly compound paths, the thin lines represent links, and τ_1 and τ_2 are allowed to be empty). The preceding notion of preemption is the usual notion of off-path preemption of Horty (1994) and Touretzky (1986). It follows from the definition that links are never preempted.

In Definition 2.3.3, if σ is preempted because $\sigma^{\mathrm{i}}\tau_1 v \tau_2 \sigma^{\mathrm{e}} \in \Phi$ (with $v \not\rightarrow x$ or $v \rightarrow x$ in Γ), then the compound path $\sigma^{\mathrm{i}}\tau_1 v \tau_2 \sigma^{\mathrm{e}}$ is called the *preempting path*; if σ is preempted because $\sigma^{\mathrm{i}} \not\rightarrow x$ or $\sigma^{\mathrm{i}} \rightarrow x$ is in Γ, then the link $\sigma^{\mathrm{i}} \not\rightarrow x$ or $\sigma^{\mathrm{i}} \rightarrow x$ itself is called the *preempting path*.

DEFINITION 2.3.4 *A set of paths Φ is* coinductive *if it is closed under initial segments, i.e., $\rho \in \Phi$ whenever $\rho x \in \Phi$.*

We are finally ready to introduce our new notion of extension. Recall that we are going to allow paths not explicitly constructed to preempt other paths from being constructed.

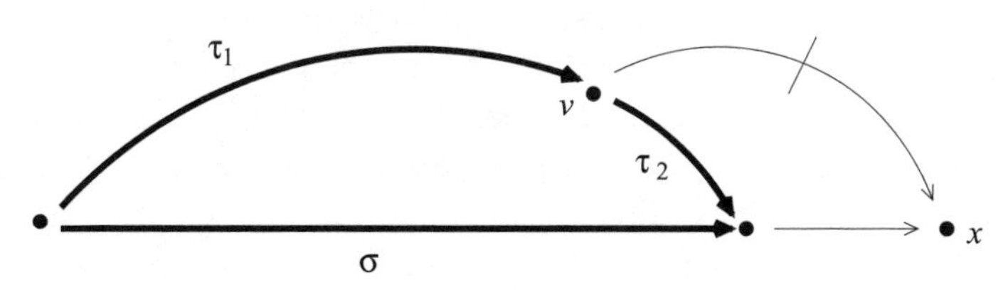

FIGURE 2.5. Preemption.

DEFINITION 2.3.5 *Let* Φ *be any coinductive set of paths. An* extension *(for* Φ*) is a pair* (Φ^+, Φ^-) *of sets of paths from* Φ *simultaneously satisfying the following two fixpoint equations:*

$$\begin{aligned}
\Phi^+ = \{\sigma \in \Phi :\ & \sigma \textit{ constructible in } \Phi^+ \ \& \\
& \sigma \textit{ not conflicted in } \Phi^+ \ \& \\
& \sigma \textit{ not preempted in } \Phi - \Phi^-\}, \\
\Phi^- = \{\tau \in \Phi :\ & \textit{some prefix of } \tau \textit{ is conflicted or preempted in } \Phi^+\}.
\end{aligned}$$

By abuse of language, a pair (Φ^+, Φ^-) *is an extension for* Γ *if and only if it is an extension for* $\Gamma^\#$.

The following theorem makes good on our promise that any defeasible inheritance network has an extension in the preceding sense.

THEOREM 2.3.1 *Every finite coinductive set* Φ *has an extension.*

For the proof of Theorem 2.3.1, see Section 2.6. Here, just the main ideas of the argument are sketched. First, we let let Γ_Φ be the set of links in Φ. We can obtain an extension for Φ in stages. At the outset, we set $\Phi_0^+ = \Gamma_\Phi$ and $\Phi_0^- = \emptyset$. For the inductive step, we let Φ_{n+1}^+ be a maximal conflict-free set of paths from Φ such that every path σ in it is (1) constructible in Φ_n^+, (2) not conflicted in Φ_n^+, and (3) not preempted in $\Phi - \Phi_n^-$. On the negative side, we define Φ_{n+1}^- as the set of all paths one of whose initial segments is conflicted or preempted in Φ_{n+1}^+. The limit of the sequence thus obtained is the desired extension.

It is important to notice how the construction brings to the foreground an asymmetry in the way conflict and preemption are handled. This asymmetry is already implicit in the notion of extension given in Definition 2.3.5: Whereas a path σ is put in Φ^+ only if it is not conflicted in Φ^+, the same path needs not to be preempted in the complement of Φ^-. In other words, based on this notion of extension, *conflicts are handled credulously*, whereas *preemption is handled skeptically*.

We also notice that the maximal set of paths mentioned in the inductive step of the preceding construction always exists. In contrast, the credulous nature of the way conflicts are handled could be mitigated somewhat by dropping this maximality clause. We would thereby obtain a more general notion, more in line with the "flexible" extensions of Horty (1994). [An even more conservative notion of extension can be obtained by proceeding similarly to Horty et al., 1990, and taking the set of all paths satisfying clauses (1)–(3) and deleting any conflicting pairs.] We will come back to these issues at the end of Subsection 2.5.2.

Let us consider the simple example of Fig. 2.1. Using the simplified node labeling t = Tweety, p = Penguin, b = Bird, f = Flies, let us see how an extension for this net can be obtained. The sets of paths obtained at each stage in the construction of Theorem 2.3.1 are as follows:

$$\begin{aligned}
\Phi_0^+ &= \{\mathtt{tp}, \mathtt{pb}, \mathtt{bf}, \mathtt{p}\overline{\mathtt{f}}\},\\
\Phi_0^- &= \emptyset,\\
\Phi_1^+ &= \Phi_0^+ \cup \{\mathtt{tpb}, \mathtt{tp}\overline{\mathtt{f}}\},\\
\Phi_1^- &= \{\mathtt{pbf}, \mathtt{tpbf}\},
\end{aligned}$$

after which a fixed point is reached. At the zeroth stage, only the links are explicitly constructed, and no paths are ruled out. At the next stage we consider all constructible paths, in this case all the length-two paths. Of these, tpb and $\mathtt{tp}\overline{\mathtt{f}}$ are explicitly constructed: Obviously neither is conflicted in Φ^+, and although the first is not preempted in the complement of Φ_0^-, the second could be preempted only by a link $x \rightarrow \mathtt{f}$, where x is more specific than p, but the only such node is t, and there is no link $\mathtt{t} \rightarrow \mathtt{f}$. Now consider Φ_1^-: It contains two paths, pbf and tpbf. Of these, the first is conflicted by the link $\mathtt{p}\overline{\mathtt{f}}$, whereas the second is preempted by the path tpb. At this point, all paths in the net have been considered, and the construction reaches a fixpoint.

As a further example, consider the net of Fig. 2.3, which has no credulous extension. The iterative process of Theorem 2.3.1 yields an extension (Φ^+, Φ^-) in which neither path $\sigma_1 = a \rightarrow b \rightarrow c$ nor $\sigma_2 = a \rightarrow b \rightarrow c \rightarrow d \rightarrow e \rightarrow b$ is in Φ^+. Clearly σ_2 cannot be preempted or conflicted (as there is no negative link incident upon b), so, a fortiori, it cannot be conflicted in Φ^+; so σ_2 is not in Φ^-. Because σ_2 preempts σ_1, it follows that σ_1 is preempted by a path in $\Phi - \Phi^-$ (for the appropriate path set Φ), and so it is not in Φ^+. Obviously, σ_2 cannot be in Φ^+, as its initial segment σ_1 is not. Also, because the only path preempting σ_1 is not in Φ^+, σ_1 is not in Φ^-. It also follows that neither path σ_1 nor σ_2 is in Φ^-. In particular, we have that σ_1 is neither in Φ^+ nor in Φ^-, and this witnesses the three-valued character of this notion of extension.

To show that this approach yields the same results as the classical one in the case of acyclic networks, we need to use the relation $\prec$, which is given in Definition 2.3.6.

Definition 2.3.6 *Let α and β be nonempty paths, with $\beta = x_0 \ldots x_n$; we say that α is* below β, *written as $\alpha \prec \beta$, if and only if $\alpha^{\mathrm{i}} = \beta^{\mathrm{i}}$, and, for some $k < n$, $\alpha^{\mathrm{e}} = \beta_k$.*

The preceding definition requires that, if $\alpha \prec \beta$, then the final node of α occurs as a node of β in a position other than the last one (although it might be repeated and occur as the last node too). The intuition behind the definition of $\prec$ is that, if $\alpha \prec \beta$, then α is a "potentially" preempting path for β (depending on whether the net contains a node v on α such that the link $v \not\Rightarrow \beta^{\mathrm{e}}$ is in Γ). Notice that if α is a proper initial segment of β then $\alpha \prec \beta$.

Recall that a relation R is well-founded over a set X if there are no infinite descending R-chains in X, or, equivalently, if any nonempty subset of X contains an R-minimal element.

Lemma 2.3.1 *Let Γ be a set of links and Φ be a coinductive set of paths over Γ. If Φ is cyclic then $\prec$ is non-well-founded on Φ.*

Proof. First observe that, if σ is a cycle, then we have $\sigma \prec \sigma$, as $\sigma^{\mathrm{i}} = \sigma^{\mathrm{i}}$ and $\sigma^{\mathrm{e}} = \sigma_k$ for some $k < \ell(\sigma)$. Similarly, if a cycle σ is a subpath of a path in Φ, we have $\tau\sigma \in \Phi$ for some τ, whence $\tau\sigma \prec \tau\sigma$. ■

Remark. Under certain conditions it is possible to reverse the implications of Lemma 2.3.1. Suppose for instance that there is a $\prec$-loop $\sigma_1 \prec \cdots \prec \sigma_n \prec \sigma_1$, there are no nodes x, y such that $x \not\Rightarrow y$ is in Γ, and y occurs on σ_i (in particular, all paths σ_i are positive). Then we can obtain a cycle as follows: We start with σ_n^{e}, which occurs on σ_1; we "follow" σ_1 to its end node σ_1^{e}, which lies on σ_2; we follow σ_2 to its end node, ...; eventually we reach σ_n, which we follow to its end node σ_n^{e}, where we started. Because there are no incident negative links, the sequence of nodes encountered is actually a path in Γ^*. This shows that if the non-well-foundedness of $\prec$ on Γ^* derives from a positive loop, then Γ cannot be acyclic. To make a similar point in a different way, if Γ is a net and Φ is the set of all trails over Γ, then $\prec$ is well-founded on Φ if and only if Γ is acyclic.

The following fact is easily established (and its proof omitted); it can be taken as further motivation for our choice of $\Gamma^{\#}$.

LEMMA 2.3.2 *Suppose* Γ *is acyclic; then* $\Gamma^* = \Gamma^\#$.

Next, we show that our notion of extension agrees with that of credulous extension in the case of acyclic graphs. The relation $\prec$ will play a crucial role in establishing this. The notion of extension subsequently introduced is classical in that it is the analog of two-valued approaches, whereas extensions in general are three-valued.

DEFINITION 2.3.7 *Let* (Φ^+, Φ^-) *be an extension for* Φ*; then* (Φ^+, Φ^-) *is* classical *if and only if* $\Phi = \Phi^+ \cup \Phi^-$.

When (Φ^+, Φ^-) is a classical extension then $\Phi^+ = \Phi - \Phi^-$, and the two fixpoint equations defining extensions collapse into one. It follows that Φ^+ is a credulous extension of Γ in the sense of Horty (1994), i.e., Φ^+ is the set of all paths that are constructible in Φ^+, but neither conflicted nor preempted in Φ^+.

THEOREM 2.3.2 *Let* Γ *be acyclic and* (Φ^+, Φ^-) *be an extension for* $\Gamma^\#$ *(which, in this case,* $= \Gamma^*$*). Then* (Φ^+, Φ^-) *is classical.*

2.4 NON-WELL-FOUNDED NETWORKS

In Section 2.3 it was shown how to obtain extensions of networks that may contain cycles, and that these extensions coincide with the credulous extensions in the case of acyclic nets. In this section we explore these extensions a little more closely and show that a network can have extensions that are pointwise $\subseteq$-smaller than one another. In contrast, acyclic networks can have only credulous extensions that are $\subseteq$-incomparable: The proof of this fact in Horty (1994) crucially employs the hypothesis of acyclicity.

DEFINITION 2.4.1 *Let* Φ *be a finite set of paths; the* well-founded part *of* Φ, $\mathsf{WF}(\Phi)$, *is defined as follows:*

$$\begin{aligned} \Psi_0 &= \emptyset; \\ \Psi_{n+1} &= \{\sigma \in \Phi : (\forall \rho \in \Phi)[\rho \prec \sigma \Rightarrow \rho \in \Psi_n]\}; \\ \mathsf{WF}(\Phi) &= \textstyle\bigcup_{n \in \mathbb{N}} \Psi_n. \end{aligned}$$

Of course, if Φ contains k paths, then $\mathsf{WF}(\Phi) = \bigcup_{n \leq k} \Psi_n$.

LEMMA 2.4.1 *Let* Φ *be a finite set of paths; then* $\prec$ *is well-founded on* Φ *if and only if* $\Phi = \mathsf{WF}(\Phi)$.

Proof. In one direction, observe that if $\rho \in \Psi_n$, then any descending $\prec$-chain from ρ is of length at most n. Consequently, if $\prec$ is not well-founded on Φ there is a path that is never put in Ψ_n, because not all of its $\prec$-predecessors are in Ψ_n. Conversely, if $\prec$ is well-founded on Φ then any descending $\prec$-chains from ρ are finite and without repetitions; moreover, because Φ is finite the length of such chains is bounded by some n. Hence, $\rho \in \Psi_{n+1}$. ■

If $\prec$ is well-founded on Φ we shall also say that Φ itself is well-founded. Similarly, if $\prec$ is not well-founded on Φ, we say that Φ itself is non-well-founded. The following theorem shows that it is the well-foundedness of $\prec$ on acyclic nets that makes the difference.

THEOREM 2.4.1 *Let Φ be a finite set of paths with the property that, if some $\rho \in \Phi$ is conflicted or preempted by $\sigma \in \Phi$, then $\rho \in \mathsf{WF}(\Phi)$ if and only if $\sigma \in \mathsf{WF}(\Phi)$. Let (Φ^+, Φ^-) be an extension for Φ, and put $\Psi^+ = \Phi^+ \cap \mathsf{WF}(\Phi)$, and similarly $\Psi^- = \Phi^- \cap \mathsf{WF}(\Phi)$. Then (Ψ^+, Ψ^-) is a classical extension for $\mathsf{WF}(\Phi)$.*

Sketch of Proof. By the hypothesis on Φ, it is immediate to verify that (Φ^+, Φ^-) is an extension for $\mathsf{WF}(\Phi)$. Arguing as in the proof for acyclic Γ, with the fact that if a path is in $\mathsf{WF}(\Phi)$ then so are its initial segments, it is possible to establish by $\prec$-induction that $[\mathsf{WF}(\Phi) - \Psi^-] \subseteq \Psi^+$. ■

The extensions we have considered so far are all *minimal*, in the sense that, of any two, neither one is extended by the other. We now look at nonminimal extensions. To do this, we begin with the following definition, which singles out certain pairs of sets of paths as being *sound*; in turn, sound pairs can be used to construct nonminimal extensions. Again this terminology is derived from the theory of truth (Kripke, 1975; Gupta and Belnap, 1993).

DEFINITION 2.4.2 *Let Φ be a set of paths closed under initial segments, and $\Psi^+, \Psi^- \subseteq \Phi$; then (Ψ^+, Ψ^-) is* sound *for Φ if the following conditions all hold:*

1. *For every path $\sigma \in \Psi^+$,*
   a. *σ is constructible in Ψ^+;*
   b. *σ not conflicted in $\Psi^+ \cup \Gamma_\Phi$, where Γ_Φ is the set of links in Φ;*
   c. *σ is not preempted in $\Phi - \Psi^-$.*

2. *Every path in Φ^- has a prefix that is either conflicted or preempted in Ψ^+.*

Obviously $(\emptyset, \emptyset)$ is sound. The point of the preceding definition is that if (Ψ^+, Ψ^-) is sound then it can be used as a starting point in the construction of an extension for Φ in such a way that no paths are lost at the next iteration. In other words, if (Υ^+, Υ^-) is obtained from (Ψ^+, Ψ^-) in the same way as (Φ_1^+, Φ_1^-) is obtained from (Φ_0^+, Φ_0^-) in Theorem 2.3.1, then $\Psi^+ \subseteq \Upsilon^+$ and $\Psi^- \subseteq \Upsilon^-$.

THEOREM 2.4.2 *Let (Ψ^+, Ψ^-) be sound for for a coinductive set Φ of paths; then there is an extension for Φ extending (Ψ^+, Ψ^-).*

Sketch of Proof. Let (Φ_n^+, Φ_n^-) be a sequence generated as in Theorem 2.3.1, except that (Ψ^+, Ψ^-) is used as a starting point, i.e.:

$$(\Phi_0^+, \Phi_0^-) = (\Psi^+, \Psi^-).$$

Show that the sequence is increasing by induction on n. The case of $n = 0$ follows immediately from the assumptions, and the case of $n > 0$ is established similarly to the corresponding case of Theorem 2.3.1. Let (Φ^+, Φ^-) be the limit of this sequence, and show that (Φ^+, Φ^-) is an extension. ■

Perhaps it is best to go through an example. Consider the net Γ of Fig. 2.6, together with the corresponding $\Gamma^\#$ (which contains 58 paths). The two paths abcd and adeb preempt each other, as there are links $\mathtt{e} \not\rightarrow \mathtt{d}$ and $\mathtt{c} \not\rightarrow \mathtt{b}$. Similarly, and for the same reasons, bcd and deb preempt each other. Moreover, the paths cdeb and adebcd are preempted by their extensions $\mathtt{cdebc}\overline{\mathtt{b}}$ and $\mathtt{adebcde}\overline{\mathtt{d}}$, respectively (and vice versa).

Now let (Ψ^+, Ψ^-) be a pair of sets of paths, where Ψ^+ contains adeb, bcd, and all the links, and Ψ^- contains abcd, deb and the paths from $\Gamma^\#$ having adeb as a prefix. The pair (Ψ^+, Ψ^-) is sound, and can be used as a starting point for the construction of an extension, which is detailed in Table 2.1. The first section of the table contains (Ψ^+, Ψ^-), and the second section gives the paths allowed or ruled at the next (and last) stage in the construction. The construction already reaches a fixed point at the second stage, giving an extension that explicitly constructs the paths in Ψ^+ and explicitly preempts the paths in Ψ^-. Obviously, such an extension is nonminimal (neither abcde nor adeb is in the well-founded part of the net, so neither one is in the least extension).

TABLE 2.1. *Construction Stages for the Net of Fig. 2.6*

+	−
adeb, bcd, ab, ad, bc, cd, de, eb, c$\bar{\text{b}}$, e$\bar{\text{d}}$	deb, debc, debcd, debcde, debcdeb, debcde$\bar{\text{d}}$, abcd, abcde, abcdeb, abcdebc, abcdebc$\bar{\text{b}}$, abcdebcd, abcde$\bar{\text{d}}$, abc$\bar{\text{b}}$, cde$\bar{\text{d}}$, ade$\bar{\text{d}}$, adebc$\bar{\text{b}}$, ebc$\bar{\text{b}}$, ebcd, ebcde, ebcdeb, ebcdebc, ebcdebc$\bar{\text{b}}$, ebce$\bar{\text{d}}$,
abc ade, cde, ebc, bc$\bar{\text{b}}$, de$\bar{\text{d}}$, adebc, bcde	bcdeb, bcdebc, bcdec$\bar{\text{b}}$, bedebcd, bcde$\bar{\text{d}}$

We can notice several features of such an extension:

1. First, the extension is nonminimal (as we noticed already), but also nonmaximal (as we could have put cdeb and adebcd in Ψ^+ and cdebc$\bar{\text{b}}$ and adebcde$\bar{\text{d}}$ in Ψ^-). Notice that although cdeb preempts cdebc$\bar{\text{b}}$ and vice versa, the situation is not completely symmetric: We could *not* have put cdebc$\bar{\text{b}}$ in Ψ^+ and cdeb in Ψ^-, for then cdebc$\bar{\text{b}}$ would not have been constructible. (The same holds, of course, for the pair adebcd and adebcde$\bar{\text{d}}$.)

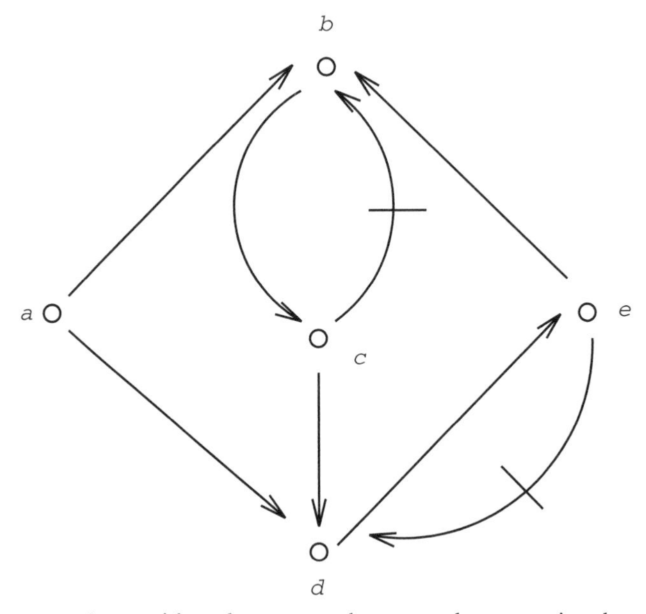

FIGURE 2.6. A net with paths abcd and adeb each preempting the other.

2. A second feature is related to the "decoupling" problem, which is considered in Subsection 2.5.1. The extension contains (by construction) the path `adeb` while ruling out the subpath `deb` (and symmetrically for `abcd` and `bcd`). The path `adeb` seems to point to the fact that `a` is a `b` in virtue of being a `d`, at the same time as the claim "`d`s are `b`s" is denied. This points to a *decoupling* of conclusions about `a`s from conclusions about `d`s. (More about this in Subsection 2.5.1.)
3. A third, perhaps undesirable, feature is the artifact of our definition. This has to do with the fact that the paths $\texttt{bc}\overline{\texttt{b}}$ and $\texttt{de}\overline{\texttt{d}}$ are allowed. Such paths seem to support conclusions to the effect that `b`s are not `b`s and that `d`s are not `d`s. This is somewhat counterintuitive, but harmless, and our definitions could easily be modified to take care of this.

We now want to look a little more closely at the ways in which the $\prec$ relation over a network might fail to be well-founded. When both Γ and Γ^* are finite, $\prec$ can be non-well-founded only if there are loops in the relation $\prec$. We single out certain specific loops in the following definitions.

Definition 2.4.3 *A finite set of paths* $\{\sigma_1, \ldots, \sigma_n\}$ *is a* $\prec$ loop *if* $\sigma_1 \prec \sigma_2 \prec \cdots \prec \sigma_n \prec \sigma_1$. *A* $\prec$*-loop of cardinality n is called an n-loop.*

Definition 2.4.4 *Let* Φ *be a set of paths; then* $\mathsf{C}(\Phi)$ *is the* closure *of* Φ *under the "initial segment" relation, i.e., the smallest set of paths having* Φ *as a subset and containing the initial segments of any path in it. Of course, by definition,* $\mathsf{C}(\Phi)$ *is coinductive.*

Definition 2.4.5 *Let* x, y *be nodes occurring on a path* σ*; we say that* x *occurs in* σ properly after y *if* σ *has the form* $\alpha y \beta x \gamma$*, for some nonempty* β.

The path $\sigma = abcdeb$ of Fig. 2.3 is a loop of cardinality 1, as σ is incident upon a node of σ in a position other than the last one. Another loop is given by the two paths *abcd* and *adeb* in Fig. 2.6, as each one of them has the same initial node as the other, but its end node lies properly on the other (i.e., lies on the other, but it is not the last node occurrence).

Definition 2.4.6 *An n-loop* $\Phi = \{\sigma_1, \ldots, \sigma_n\}$ *is* complete *with respect to* Γ *if and only if the following conditions hold:* (1) *no* σ *in* $\mathsf{C}(\Phi)$ *is conflicted in* $\mathsf{C}(\Phi)$*;* (2) *for any i such that* $0 < i < n$, σ_i *is the unique path in* $\mathsf{C}(\Phi)$

preempting an initial segment (not necessarily proper) of σ_{i+1}*; and* (3) σ_n *is the unique path* $\mathsf{C}(\Phi)$ *preempting an initial segment of* σ_1.

In other words, a loop $\{\sigma_1, \ldots, \sigma_n\}$, with $n > 1$, is complete if for every i $(0 < i < n)$ there are nodes $x_i \in \sigma_i$ and $y_i \in \sigma_{i+1}$ such that y_i occurs in σ_{i+1} properly after σ_i^{e}; and $x_i \not\rightarrow y_i$ is in Γ, if σ_{i+1} is positive, and $x_i \rightarrow y_i$ is in Γ otherwise (and similarly for σ_n and σ_1). In the special case in which $n = 1$ we say that the loop $\{\sigma_1\}$ is complete if and only if there is $\tau \sqsubset \sigma_1$ such that σ_1 preempts τ in Γ, i.e., if and only if $\sigma_1 \prec \tau$ and for some node $x \in \sigma_1$ the link $x \not\rightarrow \tau^{\mathrm{e}}$ is in Γ.

Consider again the paths *abcd* and *adeb* of Fig. 2.6. When considered in isolation, they form a complete loop, as each one of them preempts the other (relative to the net given in the figure). In particular, *abcd* preempts *adeb*, as (1) it has the same initial node as the latter, (2) is incident upon a node of the latter, viz., *d*, occupying a nonfinal position, and (3) there is a negative link $c \not\rightarrow b$. Similarly, *adeb* preempts *abcd*, as (1) it has the same initial node, (2) it is incident upon a node of the latter, viz., *b*, occupying a nonfinal position, and (3) there is a negative link $e \not\rightarrow d$. (Observe also that *abcde* and *adebc* form a loop, but one in which the uniqueness condition fails, as both *abcde* and *abcd* preempt *adeb* or one of its initial segments.)

There is a difference between complete $(2n)$-loops and complete $(2n+1)$-loops: The former, but not the latter, can be consistently partitioned into two sets Φ^+ and Φ^- that are included in some extension (for a coinductive set containing the loop as a subset), provided no preemption relations hold other than the ones explicitly mentioned in the definition of a complete loop. The point is that with $(2n)$-loops we can pick a path σ_i as a member of Φ^+: This in turn will force us to put σ_{i+1} and σ_{i-1} in Φ^-; in turn, we will have to put σ_{i+2} and σ_{i-2} in Φ^+; and so on. At the end these choices will fit in together. In the case of a $(2n+1)$-loop, however, we will find ourselves having to put the same path in Φ^+ and in Φ^-, which is of course impossible. Thus $(2n+1)$ loops cannot be partitioned by any extension. From analogy with the theory of truth, we can say that $(2n+1)$-loops behave like "liars" whereas $(2n)$-loops behave like "truth-tellers."

THEOREM 2.4.3 *Let* Γ *be a net and* $\Phi = \{\sigma_1, \ldots, \sigma_{2n}\}$ *a complete* $(2n)$ *loop over* Γ. *Then there is a classical extension for* $\mathsf{C}(\Phi)$.

On the other hand, there is no way to partition a complete $(2n+1)$-loop into two subsets, one of which contains all and only the paths that preempt a path in the other one. So we cannot obtain a sound starting point for a $(2n + 1)$-loop. Moreover, such a loop, as already mentioned, cannot intersect either Φ^+ or Φ^- if (Φ^+, Φ^-) is to be an extension: If some σ_i belongs to Φ^+ or Φ^-, then σ_{i+1} must be in Φ^- or Φ^+, respectively. Eventually, we come back full circle, having to put σ_i in Φ^- or Φ^+, respectively, which contradicts the assumption that (Φ^+, Φ^-) is an extension.

This is the reason why Horty's network in Fig. 2.3 has no creduluous extensions. On the other hand, the 2-loop in Fig. 2.6 has several extensions (in our sense), including a minimal one.

2.5 EXTENSIONS AND COMPARISONS

2.5.1 Decoupling

Decoupling is a problem, first singled out by Touretzky (1986), that arises for inheritance based on credulous extensions and in connection with the forward chaining construction of paths. Indeed, decoupling is the main reason for the introduction of double chaining (again Touretzky, 1986). To illustrate the problem, consider the net of Fig. 2.7. The net has an extension containing the path abde as well as the path $\mathtt{bc}\overline{\mathtt{e}}$. It seems that such an extension claims that a is an e in virtue of its being b, while at the same time insisting that bs are not e. Conclusions about a are not properly "coupled" with conclusions about bs.

It is easy to see that this kind of problem can be traced back to forward chaining, but only in the context of credulous extensions. In fact,

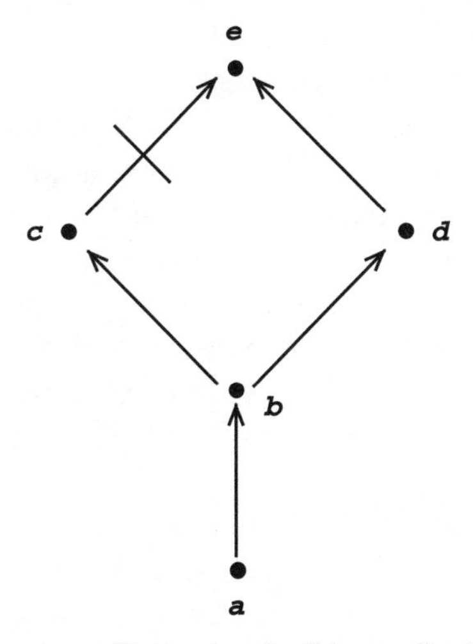

FIGURE 2.7. A net illustrating the "decoupling" problem.

one way to obviate the problem is to require that, if a path is allowed, so should all of its *terminal* segments, which would mean, of course, giving up forward chaining. But, as Makinson and Schlechta (1991) convincingly argue, forward chaining goes hand-in-hand with the basic intuition about inheritance, that specific information overrides generic information.

It seems, then, that more cautious approaches would fare better as regards the decoupling problem. In fact, the skeptical approach of Horty et al. (1990) seems to be immune to the problem. In the net of Fig. 2.7, both problematic paths abde and $\mathtt{bc\overline{e}}$ are potentially conflicted and hence not allowed.

The situation is somewhat more complicated as regards the present approach. We have already noticed how the network of Fig. 2.6 allows extensions that exhibit some form of decoupling (Section 2.4). However, such extensions are not minimal, and indeed it appears that minimal extensions would not be subject to decoupling phenomena.

2.5.2 Zombie Paths

Whereas the decoupling problem applies mostly in connection to the credulous approach, the zombie paths problem, also identified by Makinson and Schlechta (1991), applies primarily to the skeptical approaches, and it is therefore particularly relevant here.

The problem is that a skeptical approach, such as the one of Horty et al. (1990), ends up being somewhat more liberal than an approach pursuing the intersection of credulous extensions. Consider the "double diamond" of Fig. 2.8. From a skeptical point of view, neither path in the lower diamond (ats and $\mathtt{ap\overline{s}}$) is acceptable; therefore the path apq is unchallenged, and so is $\mathtt{apq\overline{r}}$. (The path atsr is not an option because its initial segment ats is unacceptable – forward chaining again.) Therefore, from a skeptical point of view, we end up endorsing $\mathtt{apq\overline{r}}$ although its potential challenger, atsr, seems intuitively to remain a genuine possibility. (Notice that a similar phenomenon is at work already in the net of Fig. 1.3 in Section 1.5, from which we conclude – skeptically – that Nixon is both a hawk and a dove, as in each case the potentially conflicting path has higher degree.)

Notice that the problem does not arise if we take the intersection of credulous extensions, for there will be an extension containing atsr and omitting $\mathtt{apq\overline{r}}$, as well as an extension containing $\mathtt{apq\overline{r}}$ and omitting atsr. Makinson and Schlechta, who first proposed the puzzle, identify

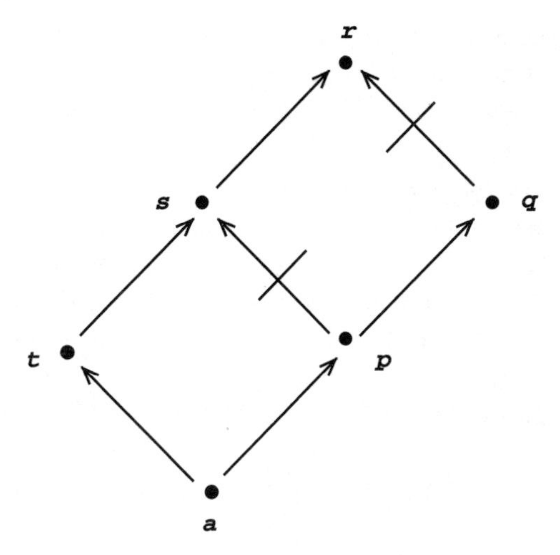

FIGURE 2.8. Zombie paths.

the root of the problem in the fact that the path `ats`, although not really a live option, must remain active enough to interfere with its competitor `apq`. It must, in other words, be "not completely dead" – a *zombie* path.

It was already mentioned that the present approach is also intended to embody the skeptical point of view, and so the question of whether it is also subject to the problem of zombie paths is relevant. As we noticed, our notion of extension handles conflicts between paths differently from the way preemption is handled: Whereas conflict is determined on the basis of the paths that are already *allowed*, preemption is (characteristically) determined on the basis of paths that are *not (yet) disallowed*. This asymmetry, reflected in the construction of Theorem 2.3.1, comes down to the fact that, although conflicts are handled credulously, preemption is handled skeptically. (Our definitions could of course be changed – in two different ways – to bring conflicts and preemption in line with each other. We do not further pursue this possibility here.)

2.5.3 Infinite Networks

At several points in this book we have used the hypothesis that the sets of paths we deal with are finite. In this section we take up the problem of how to extend the present approach to infinite sets of paths. Such an extension, although not of immediate interest for the purposes of implementation, bears some mathematical interest. The treatment in this section is meant more as an indication of a possible line of research than as a report on acquired results, and is therefore somewhat more informal than the preceding.

Infinite sets of paths can arise in a number of ways: For instance, if the net Γ contains infinitely many links, then obviously Γ^*, the set of all paths over Γ, will not be finite, independently of whether Γ itself contains cycles. Alternatively, if Γ contains cycles, then again Γ^* will not be finite, independently of whether Γ contains infinitely many links. This is why in the preceding treatment of inheritance over finite but cyclic nets we had to isolate a finite subset $\Gamma^\#$ of Γ^*, which is still, in some sense, representative of all of Γ^*.

In what follows we sketch the beginnings of a possible treatment of inheritance by using infinite sets of paths. The hypothesis that Φ is finite is used in the proof of Theorem 2.3.1 and in the definition of $\mathsf{WF}(\Phi)$ (see Definition 2.4.1). Accordingly, we indicate how to reformulate theorem and definition to take into account the possibility that Φ might now be infinite.

Given a set of paths Φ and a path $\sigma \in \Phi$, we define the rank of σ relative to Φ as follows:

$$\mathsf{rk}\,\Phi(\sigma) = \sup\{\mathsf{rk}\,\Phi(\rho) + 1 : \rho \in \Phi \wedge \rho \prec \sigma\},$$

if such a sup exists, and $\mathsf{rk}\,\Phi(\sigma) = \infty$ otherwise [in which case we say that $\mathsf{rk}\,\Phi(\sigma)$ is undefined].

Notice that, if Φ is well-founded, then $\mathsf{rk}\,\Phi(\sigma)$ exists and is defined for every $\sigma \in \Phi$, whereas if Φ is not well-founded, we will have $\mathsf{rk}\,\Phi(\sigma) = \infty$ for some $\sigma \in \Phi$. We will show that indeed there are nets containing paths of transfinite rank.

We begin by defining $\Phi_n = \{\sigma_m : m \leq n\}$, where $\sigma_m = x_1, \ldots, x_m$. Then $\mathsf{rk}\,\Phi_n(\sigma_n) = n$ for every n. Next, we remark that for any set of paths Φ it is possible to obtain a set of paths Ψ disjoint from, but topologically equivalent to, Φ, simply by renaming its nodes. So for each n, let Ψ_n be a copy of Φ_n, but with the property that Ψ_n has no nodes in common with any Ψ_m, for $m < n$. Each Ψ_n contains a path σ_n such that $\mathsf{rk}\,\Psi_n(\sigma_n) = n$.

Now put $\Psi = \bigcup_{n \geq 0} \Psi_n$. By the disjointness hypothesis, it follows that also $\mathsf{rk}\,\Psi(\sigma_n) = n$. Now we pick nodes a, b, c not occurring in Ψ, and for each n we create links $a \to \sigma_n^{\mathrm{i}}$ and $\sigma_n^{\mathrm{e}} \to b$ and refer to the path $a\sigma_n b$ as τ_n.

Then each τ_n has rank $= n$ relative to $\Psi \cup \{\tau_n : n \geq 0\}$, and moreover all the paths τ_n have the same endpoints. If we now put $\tau = abc$

and $\Omega = \Psi \cup \{\tau_n : n \geq 0\} \cup \{\tau\}$, we have $\mathsf{rk}\,\Omega(\tau) = \omega$. Notice that Ω is well-founded.

Now that we know that there are paths with infinite rank, we can extend the definition of $\mathsf{WF}(\Phi)$ to infinite sets of paths. On such sets, in general we can still define WF, but we will have to iterate the definition up the ordinal:

$$\alpha^{\Phi} = \sup\{\mathsf{rk}\,\Phi(\sigma) : \sigma \in \Phi \wedge \mathsf{rk}\,\Phi(\sigma) \neq \infty\}.$$

(It's easy to see that such an ordinal will not only be countable, as is obvious, but because $\prec$ is definable in first-order arithmetic, it will be bounded by ϵ_0. The reader is referred to Takeuti, 1987, for the details.)

In order to extend the definition of $\mathsf{WF}(\Phi)$ (Definition 2.4.1) to infinite sets of paths, we add a clause

$$\Psi_\lambda = \bigcup_{\alpha<\lambda} \Psi_\alpha,$$

for λ a limit ordinal, and define $\mathsf{WF}(\Phi)$ by taking the union over α^{Φ}. Then Lemma 2.4.1 goes through without the hypothesis on Φ.

All is left to show is how to modify the proof of Theorem 2.3.1 to allow for infinite sets of paths. First we extend the construction into the transfinite, by taking unions at limit stages, i.e., we set, for λ a limit ordinal,

$$\Phi^+_\lambda = \bigcup_{\alpha<\lambda} \Phi^+_\alpha,$$

$$\Phi^-_\lambda = \bigcup_{\alpha<\lambda} \Phi^-_\alpha.$$

We carry out the construction up to stage α^{Φ}, i.e., we set

$$\Phi^+ = \bigcup_{\alpha<\alpha^{\Phi}} \Phi^+_\alpha,$$

and similarly for Φ^-.

The argument to the effect that the sequence of sets of paths obtained in this way is increasing goes through as before. In the proof of Theorem 2.3.1 the hypothesis of finiteness is used in showing that if σ is not preempted in $\Phi - \Phi^-$ then it cannot be preempted in any $\Phi - \Phi^-_n$ for any n.

In the present context, to carry out the proof that if σ is not preempted in $\Phi - \Phi^-$ then it cannot be preempted in any $\Phi - \Phi^-_\alpha$ for any α, we proceed as follows. First we show that if $\max[\ell(\sigma), \mathsf{rk}\,\Phi(\sigma)] = \alpha$ then $\sigma \in \Phi^+$ if and only if $\sigma \in \Phi^+_\alpha$, and similarly for Φ^-: This can be shown by induction on α or, which comes to the same thing, by $\prec$-induction.

It follows that if σ were preempted by ρ in some $\Phi - \Phi^-_\alpha$ – where $\alpha = \max[\ell(\sigma), \mathsf{rk}\,\Phi(\sigma)]$ as previously – then it would be preempted in $\Phi - \Phi^-$ as well, as desired.

2.6 PROOFS OF SELECTED THEOREMS

THEOREM 2.2.1 $\mathcal{T}$ is finite.

Proof. First, observe that all branches in $\mathcal{T}$ are finite. Indeed, if N is the cardinality of C_Γ, then $N+2$ is a bound on the length of the branches in $\mathcal{T}$. So, to show that $\mathcal{T}$ is finite, it suffices to show that any sequence in the tree can have only finitely many immediate successors extending it.

Consider a sequence $\langle\lambda, \alpha_1, \ldots, \alpha_k\rangle$ in $\mathcal{T}$, and let $\rho_1 \ldots \rho_{k-1}$ be the corresponding cycles. We know that there are finitely many $\sigma \in C_\Gamma - \{\rho_1 \ldots \rho_{k-1}\}$, and for each such σ there are only finitely many τ such that $\alpha_k \leq_\sigma \tau$, as σ can be spliced in at finitely many nodes on α_k. It follows that there are only finitely many sequences $\langle\lambda, \alpha_1, \ldots, \alpha_{k+1}\rangle$ in $\mathcal{T}$.

Indeed, we can obtain a sharper bound as follows: Again consider a sequence $\langle\lambda, \alpha_1, \ldots, \alpha_k\rangle$. Let Q be the length of the longest path in S_Γ and P the length of the longest cycle in C_Γ. Then for any $k \leq N+2$ the length of α_k is bounded by $Q+kP$. Moreover, given a sequence such as the preceding one, there are only $card\,(C_\Gamma) - k$ cycles available for splicing, and each can be spliced into at most $Q+kP$ nodes. If we now put

$$X = \prod_{k=0}^{N+2} [card\,(C_\Gamma) - k](Q+kP),$$

then the branching factor is bounded by $\max[X, card\,(S_\Gamma)]$. ∎

THEOREM 2.3.1 *Every finite coinductive set Φ has an extension.*

Proof. Assume that Φ is a finite and coinductive set of paths. (In Subsection 2.5.3 the assumption of finiteness is dropped.) Let Γ_Φ be the set of links in Φ, i.e., the set of length-one paths in Φ. We construct an extension for Φ in stages. For every $n \geq 0$ we define Φ^+_n and Φ^-_n inductively as follows:

- $\Phi^+_0 = \Gamma_\Phi$ and $\Phi^-_0 = \emptyset$;
- $\Phi^+_{n+1} =$ a maximal conflict-free set of paths σ from Φ such that the following all hold:

1. σ is constructible in Φ_n^+;
2. σ is not conflicted in Φ_n^+;
3. σ is not preempted in $\Phi - \Phi_n^-$.

- $\Phi_{n+1}^- = \{\tau \in \Phi : \text{some prefix of } \tau \text{ is conflicted or preempted in } \Phi_{n+1}^+\}$.

We then set

$$\Phi^+ = \bigcup_{n \in \mathrm{N}} \Phi_n^+,$$

$$\Phi^- = \bigcup_{n \in \mathrm{N}} \Phi_n^-.$$

We now show, by induction on n, that $\Phi_n^+ \subseteq \Phi_{n+1}^+$ and $\Phi_n^- \subseteq \Phi_{n+1}^-$.

BASE CASE $n = 0$. A link $x \to y$ or $x \not\to y$ in Γ_Φ is constructible, never conflicted, and never preempted; so $x \to y$ or $x \not\to y$, respectively, is in Φ_1^+. So $\Phi_0^+ = \Gamma_\Phi \subseteq \Phi_1^+$. Moreover, $\Phi_0^- = \emptyset \subseteq \Phi_1^-$.

INDUCTIVE STEP. Suppose $\sigma \in \Phi_{n+1}^+$. If σ is a link, $\sigma \in \Phi_{n+2}^+$, as before. If σ is not a link, then $\ell(\sigma) \geq 2$, and say $\sigma = \rho x$, with $\rho \in \Phi_n^+$ and $\rho^{\mathrm{e}} \to x \in \Gamma_\Phi$ (the case in which σ is negative is similar). By the inductive hypothesis, we have $\rho \in \Phi_{n+1}^+$, which gives that σ is constructible in Φ_{n+2}^+ as well.

Now we show that σ is not preempted in $\Phi - \Phi_{n+1}^-$. Therefore we assume by way of contradiction that σ is preempted in $\Phi - \Phi_{n+1}^-$. This means that there is a node v such that $v \not\to x$ is in Γ, and either $v = \rho^{\mathrm{i}}$ or $\rho^{\mathrm{i}}\tau_1 v \tau_2 \rho^{\mathrm{e}} \in \Phi - \Phi_{n+1}^-$ for some paths τ_1 and τ_2. So we further distinguish two cases. If $v = \rho^{\mathrm{i}}$, then $\rho^{\mathrm{i}} \not\to x$ is in Γ, which would make σ already preempted in $\Phi - \Phi_n^-$, as Φ_n^- contains no direct links. The other case is $\rho^{\mathrm{i}}\tau_1 v \tau_2 \rho^{\mathrm{e}} \in \Phi - \Phi_{n+1}^-$. By the inductive hypothesis, $\Phi_n^- \subseteq \Phi_{n+1}^-$, whence $\rho^{\mathrm{i}}\tau_1 v \tau_2 \rho^{\mathrm{e}} \in \Phi - \Phi_n^-$. But this makes σ preempted in $\Phi - \Phi_n^-$, whence $\sigma \notin \Phi_{n+1}^+$, against assumption. This shows that $\sigma \in \Phi_{n+2}^+$.

Finally, σ is not conflicted in Φ_{n+1}^+, as Φ_{n+1}^+ is conflict free. So, if still $\sigma \notin \Phi_{n+2}^+$, it is because it is conflicted in Φ_{n+2}^+. Let τ be the conflicting path: Then τ is not a link (or else $\sigma \notin \Phi_{n+1}^+$). It follows that τ is conflicted in Φ_{n+1}^+ (by σ), which is impossible if τ is in Φ_{n+2}^+.

Now suppose $\sigma \in \Phi_{n+1}^-$: So σ extends some τ that is conflicted or preempted in Φ_{n+1}^+; by the inductive case for Φ_{n+1}^+, the conflicting or preempting path is also in Φ_{n+2}^+, whence $\sigma \in \Phi_{n+2}^-$, as required.

We now show that (Φ^+, Φ^-) is an extension. Although this is an immediate consequence of the finiteness of Φ, the following is a more general argument, which will be used subsequently. We need to establish, first, that if σ is constructible in Φ^+, but neither conflicted in Φ^+ nor preempted in $\Phi - \Phi^-$, then $\sigma \in \Phi^+$. Therefore suppose σ is constructible in Φ^+ but neither conflicted in Φ^+ nor preempted in $\Phi - \Phi^-$. Then (1) there is a stage p such that σ is constructible in all the Φ^+_m for $m \geq p$ (this follows from the fact that the sequence we construct is increasing and Φ^+ is its limit); (2) at no stage n can σ be conflicted in Φ^+_n, or else it would be conflicted in the limit too; (3) suppose that for every n, σ is preempted in $\Phi - \Phi^-_n$, by some preempting path ρ_n. Because Φ is finite and $\Phi - \Phi^-_n$ *decreases* as n increases, there is some ρ_{n_0} such that ρ_{n_0} preempts σ and

$$\rho_{n_0} \in \Phi - \bigcup_{n \geq 0} \Phi^-_n = \Phi - \Phi^-,$$

which is impossible. It follows that σ is preempted in $\Phi - \Phi^-$ too, against assumption. So there is q such that, for every $n \geq q$, σ is not preempted in $\Phi - \Phi^-_n$. Now let $n^* = \max(p, q)$. Then σ is constructible in $\Phi^+_{n^*}$ by (1), but neither conflicted in $\Phi^+_{n^*}$ nor preempted in $\Phi - \Phi^-_{n^*}$, by (2) and (3): so $\sigma \in \Phi^+_{n^*+1} \subseteq \Phi^+$. Likewise, if some prefix of σ is conflicted or preempted in Φ^+, we obtain $\sigma \in \Phi^-$.

Conversely, we need to show that if $\sigma \in \Phi^+$ then any prefix of σ is constructible in Φ^+, but neither conflicted in Φ^+ nor preempted in $\Phi - \Phi^-$; so we let $\sigma \in \Phi^+$. If σ were conflicted by some $\rho \in \Phi^+$, then there would be n such that both $\sigma, \rho \in \Phi^+_n$, which is impossible, as all Φ^+_n are conflict free by construction. Similarly, if σ were not constructible in Φ^+, then it would not be constructible in any Φ^+_n, and hence it could not be in Φ^+. Finally, if σ were preempted by $\rho \in \Phi - \Phi^- = \bigcap_{n \geq 0}(\Phi - \Phi^-_n)$, then for every n we would have $\rho \in \Phi - \Phi^-_n$; then σ would be preempted in $\Phi - \Phi^-_n$ for every n, and hence never put into Φ^+, against hypothesis.

Finally, we need to show that $\sigma \in \Phi^-$ if and only if some prefix τ of σ is preempted or conflicted in Φ^+. We have $\sigma \in \Phi^-$, iff for some n, $\sigma \in \Phi^-_n$; iff, by construction, σ is preempted or conflicted in Φ^+_n, and iff σ is also preempted or conflicted in Φ^+ as well.

This shows that (Φ^+, Φ^-) is an extension for Φ. Moreover, given that Φ is finite (as when, e.g., $\Phi = \Gamma^\#$) then because Φ^+_n and Φ^-_n are subsets of Φ, this extension is reached at some finite stage. ■

THEOREM 2.3.2 Let Γ be acyclic and (Φ^+, Φ^-) be an extension for $\Gamma^\#$ (which, in this case, $= \Gamma^*$). Then (Φ^+, Φ^-) is classical.

Proof. We need to show that $\Gamma^\# = \Phi^+ \cup \Phi^-$. Because Φ^+ and Φ^- are subsets of $\Gamma^\#$, it suffices to show that $\Gamma^\# - \Phi^- \subseteq \Phi^+$.

So we assume that $\sigma \in \Gamma^{\#} - \Phi^{-}$, to show that $\sigma \in \Phi^{+}$. We proceed by induction on $\prec$. Suppose that $\sigma \in \Gamma^{\#} - \Phi^{-}$, and assume that the property holds for any $\rho \prec \sigma$.

If any proper initial segment ρ of σ is in Φ^{-} then so is σ; consequently all proper initial segments of σ are in $\Gamma^{\#} - \Phi^{-}$. Because any initial segments of σ are $\prec$ below σ, it follows by the inductive hypothesis that all initial segments are in Φ^{+}. Hence σ is constructible in Φ^{+}. Moreover, if σ were conflicted in Φ^{+} then $\sigma \in \Phi^{-}$, against the hypothesis.

It remains to show that σ is not preempted in $\Gamma^{\#} - \Phi^{-}$. So assume by way of contradiction that σ is preempted in $\Gamma^{\#} - \Phi^{-}$, and let ρ be the preempting path. Then $\rho \prec \sigma$, and by inductive hypothesis $\rho \in \Phi^{+}$. This makes σ preempted in Φ^{+}, whence $\sigma \in \Phi^{-}$, against the hypothesis.

Because σ is constructible in Φ^{+}, but neither conflicted in Φ^{+} nor preempted in $\Gamma^{\#} - \Phi^{-}$, it follows that $\sigma \in \Phi^{+}$, as required. ■

THEOREM 2.4.3 Let Γ be a net and $\Phi = \{\sigma_1, \ldots, \sigma_{2n}\}$ be a complete $(2n)$-loop over Γ. Then there is a classical extension for $\mathsf{C}(\Phi)$.

Proof. Let $\Psi^{\clubsuit} = \{\sigma_{2i+1} : i < n\}$ and $\Psi^{\spadesuit} = \{\sigma_{2i+2} : i < n\}$. Now we define a starting point (Φ_0^{+}, Φ_0^{-}) by setting, say, $\Phi_0^{+} = \mathsf{C}(\Psi^{\clubsuit})$ and $\Phi_0^{-} = \{\tau : \tau$ extends a path in $\Psi^{\spadesuit}\}$. (Here $\Psi^{\spadesuit}$ and $\Psi^{\clubsuit}$ could have been switched.)

We check that (Φ_0^{+}, Φ_0^{-}) is sound for $\mathsf{C}(\Phi)$, in the sense of Definition 2.4.2. Let $\tau \in \Phi_0^{+}$, and say that τ is an initial segment of σ_{2i+1}. Obviously, τ is constructible in Φ_0^{+}. Because Φ is complete, τ is not conflicted. It remains to show that τ is not preempted in $\Phi - \Phi_0^{-}$.

Assume by way of contradiction that τ is preempted in $\Phi - \Phi_0^{-}$. If $i > 0$, then the preempting path must be σ_{2i}, and if $i = 0$ then the preempting path is σ_{2n}. In either case the preempting path is not in $\Phi - \Phi_0^{-}$.

Now let $\tau \in \Phi_0^{-}$; then τ extends σ_{2i+2} for some $i < n$. We need to show that τ has a prefix that is either conflicted or preempted in Φ_0^{+}. But this follows immediately, as τ extends σ_{2i+2}, which is preempted by $\sigma_{2i+1} \in \Phi_0^{+}$.

So (Φ_0^{+}, Φ_0^{-}) is sound for $\mathsf{C}(\Phi)$. By Theorem 2.4.2 there is an extension (Φ^{+}, Φ^{-}) extending (Φ_0^{+}, Φ_0^{-}). We need to check that such an extension is classical for $\mathsf{C}(\Phi)$.

Let τ be a path in $\mathsf{C}(\Phi)$ but not in Φ^{-}. If τ is in Φ then $\tau = \sigma_k$ for some positive $k \leq 2n$, and given that τ is not in Φ^{-} then k must be odd, so that $\tau \in \Phi_0^{+} \subseteq \Phi^{+}$. If τ is not in Φ, it is a proper initial segment of a path in Φ, and we further distinguish the following cases:

1. $\tau \sqsubset \sigma_{2i+1}$ for some $i < n$. Then again $\tau \in \Phi^+$ by construction.
2. $\tau \sqsubset \sigma_{2i+2}$ for some $i < n$. We show by induction on length that any $\tau \sqsubset \sigma_{2i+2}$ such that $\tau \notin \Phi^-$ is in Φ^+. Because (Φ^+, Φ^-) is an extension, all links are in Φ^+, which takes care of the base case. For the inductive step: If some prefix of τ were in Φ^-, then so would τ; it follows that every prefix of τ is outside Φ^- and hence by inductive hypothesis in Φ^+; so τ is constructible in Φ^+. Moreover, because of the completeness hypothesis, τ cannot be conflicted in Φ^+. Finally, if τ were preempted, given the completeness hypothesis and the fact that $\tau \sqsubset \sigma_{2i+2}$, it would have to be preempted by $\sigma_{2i+1} \in \Phi^+$, which would put τ in Φ^-, against assumption.

Thus τ is constructible in Φ^+, but neither conflicted in Φ^+ nor preempted in $\mathsf{C}(\Phi) - \Phi^-$. Because (Φ^+, Φ^-) is an extension, $\tau \in \Phi^+$, as required. ∎

3

General Extensions for Default Logic

3.1 INTRODUCTORY REMARKS

In this chapter we further develop the basic three-valued approach to defeasible inference by introducing a notion of "general extension" for Default Logic that falls in line with the analogous notion for nonmonotonic networks introduced in Chap. 2. To an extent, the present work can be viewed as a vindication of the skeptical approach to defeasible inheritance of Horty et al., without at the same time underestimating its indebtedness to the results of Fitting (1990, 1993), Przymusiński (1991), and Przymusińska and Przymusiński (1994) in logic programming.

As will be clear, such a notion of general extension provides a generalization of the notion of extension for Default Logic that was originally due to Reiter (1980). The two versions differ mainly in the way conflicts among defaults are handled. Characteristically, the present notion of general extension allows defaults not explicitly triggered to preempt other defaults. A consequence of the adoption of such a notion of extension is that the resulting framework turns out to be mathematically well behaved, in that the collection of all the general extensions of a default theory has a nontrivial algebraic structure. This fact has two major technical fallouts: First, it turns out that every default theory has a general extension; second, general extensions allow one to define a well-behaved, skeptical relation of defeasible consequence for default theories, satisfying the principles of Reflexivity, Cut, and Cautious Monotonicity formulated by Gabbay (1985) (see Section 1.2).

All nonmonotonic logics aim to provide a formal account of the fact that reasoners can reach conclusions tentatively, reserving the right to retract them in the light of further information. As a consequence, a certain amount of "jumping to the conclusion" is built right into any nonmonotonic framework, in the form (particularly explicit in the case of Default Logic) of *defeasible inference rules*. Nonmonotonicity amounts precisely to the proviso that if conflicts were to arise, one would "retract" some of the conclusions in order to restore consistency.

However, we saw in Section 1.4 that there are in principle two different kinds of conflict that can arise: (1) conflicts between tentatively endorsed conclusions and newly learned facts (strict information) about the world, and (2) conflicts among the conclusions of the defeasible rules. All nonmonotonic frameworks handle the first kind of conflict by not triggering rules whose conclusions are inconsistent with the given information; hence, if new facts were to be learned, old conclusions would be retracted (this is the essence of nonmonotonicity). On the other hand, conflicts of the second kind can be handled either *credulously* or *skeptically*. The well-known "Nixon diamond" of Fig. 1.2 (Section 1.4) was used to clarify the distinction: In the presence of two conflicting defeasible inferences (and in the absence of other considerations such as specificity), the credulous approach would have us commit to one or the other conclusion whereas the skeptical approach would have us withhold commitment.

This chapter introduces a notion of general extension for Default Logic that tries to capture some of the intuitions behind skepticism. Whereas conflicts with strict information are handled in the usual manner (thereby retaining the fundamental *desideratum* of nonmonotonicity), interactions among defeasible rules are handled somewhat differently. Just as in the case of nonmonotonic networks of Chap. 2, direct conflicts between the conclusions of different defaults are handled credulously, whereas preemption relations (suitably defined) are handled skeptically (Section 4.2 shows how this asymmetry can be overcome). As a consequence, in the case of preemption, default rules need not be explicitly triggered in order to prevent the conclusions of conflicting rules from being endorsed.

It was mentioned that, just as in the case of nonmonotonic networks, every default theory turns out to have a general extension. In this, Default Logic as based on general extensions differs from Reiter's original proposal (1980) and is more in line with Lukaszewicz (1988) and Brewka (1991), both subsumed by Delgrande, Schaub, and Jackson (1994). In fact, Delgrande et al. explicitly claim their approach to provide an "amalgamation" of those of Lukaszewicz and Brewka (1994, p. 198). The basic intuition behind general extensions is, however, quite different from these other proposals. (Section 4.3 contains a comparison of Default

Logic based on general extension to some of the other approaches in the literature.)

Moreover, general extensions allow the identification of a well-behaved relation of defeasible consequence for default theories. In Section 1.2 we saw that there are several crucial properties of a defeasible consequence relation, perhaps the most significant of which, as well as the most elusive, is Cautious Monotony. As borne out by the analysis of Makinson (1994), there seems to be no natural way of using the notion of extension of Reiter (1980) to define a relation of defeasible consequence for default logic (skeptical or otherwise), satisfying Gabbay's principles, including Cautious Monotony. Although this will be the topic of Chap. 4, it is important to keep this fact in mind as we go through the details of general extensions.

The most natural way to obtain a relation of defeasible consequence is through the already mentioned definition of a relation $\mid\!\sim$ obtaining between a default theory and a proposition φ precisely when φ is supported by *all* extensions of the theory. In other words, φ would be defeasibly inferable from a given default theory if it occurs in the intersection of (the deductive closures of) all its extensions. However, even discounting feasibility worries and other conceptual issues, as we saw in Section 1.3 there appears to be a major problem with this approach. Makinson (1994, p. 60) shows that, by defining $\mid\!\sim$ as the intersection of all the extensions (in Reiter's sense) of a default theory, one cannot satisfy Cautious Monotony. This is a substantial drawback.

It is then significant that using general extensions, the relation $\mid\!\sim$ defined as the intersection of the extensions turns out to have all three properties identified by D. Gabbay. Moreover, in the important special case of seminormal default theories, we have the existence of a unique privileged extension of the theory (the *least* extension), which makes the definition of $\mid\!\sim$ particularly simple. Admittedly, this can have, at least in some cases, some counterintuitive consequences (see Section 3.5). Although it is argued that, once the cautious motivation of the theory is properly taken into account, such consequences are less counterintuitive than it might at first appear, alternative ways of identifying a privileged extension for a default theory in such a way as to obviate some of the counterintuitive consequences are also reviewed.

The chapter is organized as follows. In Section 3.2 the notion of general extension for the particularly simple, but already significative, case of categorical (i.e., prerequisite-free) default theories is illustrated. In Section 3.3 we take up several examples of categorical theories. In Section 3.4 the notion of general extension for arbitrary default theories is introduced, and some examples are presented in Section 3.5. Section 3.6 contains proofs of selected theorems.

3.2 CATEGORICAL DEFAULT THEORIES

Fix a propositional language $\mathcal{L}$, obtained from an infinite list of propositional variables $p, q, r, \ldots$, by the propositional connectives $\wedge, \vee, \neg, \rightarrow$. The language $\mathcal{L}$ is endowed with the usual (i.e., two-valued) semantics. In particular, we use $\models$ to denote the usual relation of logical consequence. If S and W are sets of propositional axioms from $\mathcal{L}$ and φ a formula from $\mathcal{L}$, then we write $S \models_W \varphi$ as shorthand for $S \cup W \models \varphi$. A set S of sentences is W-consistent if and only if $S \cup W$ is consistent.

A *default theory* is a pair (W, Δ), where W is a set of propositional axioms (a *world description*) and Δ is a *finite* set of defaults. In turn, a *default* δ is an expression of the form

$$\frac{\zeta : \eta}{\theta},$$

where ζ, η, and θ are sentences from $\mathcal{L}$. The intuitive meaning of a default δ is that *if ζ has been established, and $\neg\eta$ has not been established, then assume θ ("by default")*. The expressions ζ, η, and θ are called the *prerequisite*, the *justification*, and the *conclusion* of δ, respectively. If $\delta = (\zeta : \eta)/\theta$, then we put $P(\delta) = \zeta$, $J(\delta) = \eta$, and $C(\delta) = \theta$. If Γ is a set of defaults, we write $C(\Gamma)$ for $\{C(\delta) : \delta \in \Gamma\}$, and similarly for $P(\Gamma)$ and $J(\Gamma)$.

In the particular case in which a default δ has no prerequisite [because it is of the form $:\eta/\theta$ or, equivalently, because $P(\delta)$ is a tautology], then we say that δ is *categorical*. If all defaults in Δ are categorical, then we say that (W, Δ) is categorical. As we will see, categorical default theories form a natural and well-behaved class. Other special cases: A default δ is *normal* if $J(\delta)$ and $C(\delta)$ are logically equivalent, and it is *seminormal* if $J(\delta)$ logically implies $C(\delta)$ [typically, because $J(\delta)$ is a conjunction one of whose conjuncts is $C(\delta)$].

DEFINITION 3.2.1 *Let S be a set of $\mathcal{L}$-sentences, W a world-description, and δ a default. Then we say that*

1. *δ is* admissible *in S (relative to W) if and only if $S \models_W P(\delta)$,*
2. *δ is* conflicted *in S (relative to W) if and only if $S \models_W \neg C(\delta)$,*
3. *δ is* preempted *in S (relative to W) if and only if $S \models_W \neg J(\delta)$.*

If Γ is a set of defaults, we say that δ is admissible, conflicted, or preempted in Γ according as δ is admissible, preempted, or conflicted in $C(\Gamma)$.

Reiter defines the notion of an extension for a default theory. Intuitively, an extension for a default theory (W, Δ) is a deductively closed, consistent set of formulas containing W and the consequents of a maximal subset of Δ. Here the slightly simpler notion of a *classical extension* for a default theory is introduced. Similar simplifications have been advocated in Papadimitriou and Sideri (1994), Brass (1993), and Dix (1992).

Definition 3.2.2 *A set Γ of defaults is a* classical extension *for (W, Δ) if and only if it satisfies $\Gamma = \bigcup_{n \geq 0} \Gamma_n$, where $\Gamma_0 = \emptyset$, and*

$$\Gamma_{n+1} = \{\delta \in \Delta : C(\Gamma_n) \models_W P(\delta) \;\&\; C(\Gamma) \not\models_W \neg J(\delta)\}$$

(notice the occurrence of Γ in the definition of Γ_{n+1}). In other words, Γ_{n+1} is the set of defaults admissible in Γ_n that are not preempted in Γ.

It follows from the definition that $C(\Gamma)$ is conflict free, because if Γ contained defaults with conflicting conclusions (say θ and $\neg\theta$), then any default would be preempted [because $\theta, \neg\theta \models \neg P(\delta)$ no matter what $P(\delta)$ is]. Observe also that if Γ is a classical extension for (W, Δ) then $\{\varphi : C(\Gamma) \models_W \varphi\}$ is an extension in Reiter's original sense. In general, a default theory might have zero, one, or more than one extension.

We record here the following well-known facts about classical extensions.

Theorem 3.2.1 *Let Γ be a classical extension for (W, Δ).*

1. *$C(\Gamma)$ is W inconsistent if and only if W is inconsistent.*
2. *If δ is admissible in Γ, but neither conflicted nor preempted in Γ, then $\delta \in \Gamma$.*

We now proceed with the promised generalization of the notion of extension, beginning with the somewhat simpler particular case of categorical theories. Let (W, Δ) be a categorical default theory, i.e., a theory whose defaults have no prerequisite. In the general case of a default theory, extensions have to be defined in the somewhat peculiar form of Definition 3.4.3 because extensions have to be "grounded" (in the sense of Section 3.4: A classical extension Γ is grounded if it is a *minimal* set of defaults admissible but not preempted in Γ). Definition 3.4.3 is formulated in such a way as to enforce this form of minimality by letting no more defaults become admissible than "have to." But when a theory is categorical, groundedness is no longer a concern, and it is immediate to check that Γ is a classical extension for a categorical theory (W, Δ) if and only if it is a solution to the following fixpoint equation:

$$\Gamma = \{\delta \in \Delta : C(\Gamma) \not\models_W \neg J(\delta)\}.$$

In other words, Γ contains all and only those defaults that are not preempted in Γ. Notice that this also forces extensions to be *conflict free*, for if $C(\Gamma)$ were conflicted, then $C(\Gamma) \models_W \theta$ for any θ and in particular $C(\Gamma) \models_W \neg J(\delta)$.

DEFINITION 3.2.3 *A* general extension *for a categorical default theory* (W, Δ) *is a pair* (Γ^+, Γ^-) *of disjoint sets of defaults from* Δ*, simultaneously satisfying the following two fixpoint equations:*

$$\Gamma^+ = \{\delta : C(\Gamma^+) \not\models_W \neg C(\delta) \;\&\; C(\Delta - \Gamma^-) \not\models_W \neg J(\delta)\};$$
$$\Gamma^- = \{\delta : C(\Gamma^+) \models_W \neg C(\delta) \;\; or \;\; C(\Gamma^+) \models_W \neg J(\delta)\}.$$

In other words, Γ^+ *is the set of all defaults that are neither conflicted in* Γ^+ *nor preempted in* $(\Delta - \Gamma^-)$*, and* Γ^- *is the set of all defaults that are either conflicted or preempted in* Γ^+*.*

Observe that the components Γ^+ and Γ^- of the extension are required to be disjoint. This corresponds to the intuition that Γ^+ is the set of the (definitely) triggered defaults and Γ^- is the set of the (definitely) excluded defaults, and that no default can be both explicitly triggered and explicitly ruled out. From a purely technical point of view, it is possible to develop the theory without such a disjointness requirement, and this would correspond to a "four-valued" intuition such as that of Belnap (1977). This possibility is not pursued here.

In the next two theorems we now show that general extensions indeed generalize the notion of classical extension and that general extensions always exist. Proofs are given in Section 3.6.

THEOREM 3.2.2 *Let* Γ^+ *be a classical extension for a categorical default theory* (W, Δ)*, and put*

$$\Gamma^- = \{\delta : \; \delta \textit{ conflicted or preempted in } \Gamma^+\};$$

then (Γ^+, Γ^-) *is a general extension for* (W, Δ)*.*

THEOREM 3.2.3 *Every categorical default theory has a general extension.*

As mentioned, full proofs of the theorems can be found in Section 3.6. It is, however, interesting intuitively to characterize the process by which such an extension can be obtained. A general extension (Γ^+, Γ^-) can be constructed "from below" in stages, as the limit of the sequences Γ_n^+ and Γ_n^- (for $n \geq 0$). For the starting point, we put $\Gamma_0^+ = \Gamma_0^- = \emptyset$.

For the inductive step, let Γ_{n+1}^+ be a *maximal* set of defaults from Δ such that (1) the conclusions of defaults in $\Gamma_n^+ \cup \Gamma_{n+1}^+$ form a consistent set, and (2) no default $\delta \in \Gamma_{n+1}^+$ is preempted in $\Delta - \Gamma_n^-$. Also put

$$\Gamma_{n+1}^- = \{\delta : \delta \text{ conflicted or preempted in } \Gamma_{n+1}^+\}.$$

The sequences are increasing and disjoint, and their limit gives a general extension.

Observe that in the defintion of Γ^+ the maximal subset need not be unique, so that the process is not deterministic (this will be important for the definition of the relation $\mid\!\sim$), and also that the definition provides further evidence for the fact that conflicts are handled credulously whereas preemptions are handled skeptically.

The existence of extensions is a desirable feature that is missing in Reiter's original formulation, but that can be found for instance in all three proposals of Lukaszewicz (1988), Brewka (1991), and Delgrande et al. (1994). But as the examples discussed in Section 3.3 will make clear, the intuitions at the basis of these proposals are different from the present one, and the formal mathematical properties of the frameworks are correspondingly different.

3.3 EXAMPLES

In this section we take up a few examples in order to compare the notion of general extension given here to other notions available in the literature. As before, "classical extension" here refers to the notion of extension in the sense of Reiter (1980), as simplified in Definition 3.2.2.

EXAMPLE 3.3.1 *Consider first the default theory* (W, Δ), *in which* W *is empty and* Δ *comprises the two defaults*

$$\frac{:p}{\neg q}, \quad \frac{:q}{\neg p}.$$

In this theory, each default preempts the other one, and, as is well known, the theory has two classical extensions, according to which default is triggered. But the theory of general extensions handles preemption skeptically, and hence in addition to these two classical extensions, the theory has one general extension, in which no default is triggered and none is ruled out.

EXAMPLE 3.3.2 *Consider the default theory in which W is empty and Δ comprises only the default*

$$\frac{:p}{\neg p}.$$

This is a similar case, in that the single default preempts itself. The theory has no classical extensions, but it has one general extension, namely $(\emptyset, \emptyset)$, in which the default is neither triggered nor ruled out (again because preemption is handled skeptically). The theory would still have the same unique general extension even if W contained p (for then the default would be conflicted) and also if it contained $\neg p$ (for then the default would be preempted).

EXAMPLE 3.3.3 *Consider the default theory in which W is empty and Δ comprises the two defaults*

$$\frac{:p}{q}, \quad \frac{:p}{\neg q}.$$

This example illustrates the way conflicts (as opposed to preemptions) are handled. Neither default preempts the other, but they are in conflict. Here, we have the analog of the Nixon diamond of Fig. 1.2 in Section 1.4. Because conflicts are handled credulously in the theory of general extensions, the general extensions of the theory coincide with the classical ones.

EXAMPLE 3.3.4 *Consider the default theory in which W is empty and Δ comprises only the default*

$$\frac{:q}{p \wedge \neg p}.$$

Whereas the single default of Example 3.3.2 was self-preempting, the single default of Example 3.3.4 is self-conflicting. Notice that the default is also self-preempting, in that $p \wedge \neg p$ implies $\neg q$, but this is an artifact of the underlying relation $\models$ of logical consequence. It's easy to see that even discounting the effects of $\models$ (i.e., even ignoring the fact that $p \wedge \neg p \models \neg q$), the theory would still have no classical extensions. But in spite of the credulous handling of conflicts, the three-valued nature of general extension still ensures the existence of an extension in which the default is neither triggered nor ruled out.

EXAMPLE 3.3.5 *Consider now the theory in which W is empty but* Δ *comprises the defaults*

$$\frac{:p}{\neg p}, \qquad \frac{:q}{r}.$$

As before, this theory has no classical extensions because of the first default. It does have one general extension, namely $(:q/r, \emptyset)$. The first default cannot be triggered, but there is no obstacle that prevents triggering the second. This testifies to the extent to which the theory of general extensions manages to keep conflicts and preemptions *local.*

EXAMPLE 3.3.6 *Consider a theory in which W is empty and* $\Delta = \{\delta_1, \ldots, \delta_n\}$ *for some* $n > 1$. *Suppose also that, for all k such that* $1 \leq k < n$,

$$\delta_k = \frac{:p_k}{\neg p_{k+1}},$$

whereas

$$\delta_n = \frac{:p_n}{\neg p_1}.$$

So defaults are organized in a "loop," in which the conclusion of each one of them preempts the next default, and the conclusion of the last one preempts the first. It is easy to check that this theory has no classical extensions if n is odd. [It has, of course, one general extension even when n is odd, namely $(\emptyset, \emptyset)$.] On the other hand, if n is even, say $n = 2m$, beside the general extension $(\emptyset, \emptyset)$, there are two classical extensions, namely

$$(\{\delta_{2k+1} : 0 \leq k < m\}, \{\delta_{2k+2} : 0 \leq k < m\}),$$

and the one with the two components switched around,

$$(\{\delta_{2k+2} : 0 \leq k < m\}, \{\delta_{2k+1} : 0 \leq k < m\}).$$

Indeed, this result is consistent with similar results in Antonelli (1997) and Papadimitriou and Sideri (1994).

So far we have been concerned with only the existence of extensions, but this is by no means the only question that can be asked when a notion of extension is proposed. In the remaining part of this section we take up a few "benchmark" examples, mostly drawn from Delgrande et al. (1994), in order to assess the behavior of the proposed notion.

EXAMPLE 3.3.7 *(Semimonotonicity.) Consider the default theory* (W_1, Δ_1), *in which* $W_1 = \emptyset$ *and* Δ_1 *comprises*

$$\frac{: p \wedge \neg q}{p}.$$

We obtain theory (W_2, Δ_2) *by putting* $W_2 = W_1$, *and by adding to* Δ_1 *the default* $: q/q$.

Semimonotonicity is a property that was first singled out by Reiter. According to this property the extensions of a theory increase monotonically with the size of Δ: If Γ is an extension for (W, Δ) and $\Delta \subseteq \Delta'$, then there is an extension Γ' for (W, Δ') such that $\Gamma \subseteq \Gamma'$.

By considering Example 3.3.7, we see that semimonotonicity fails for the notion of general extension of Section 3.2. The theory (W_1, Δ_1) has one general extension, triggering its unique default. When the second default is thrown in, however, we lose the first one, in the sense that the unique extension of the theory triggers the second default and rules out the first (the first default is now potentially preempted). In this, the version of Default Logic presented here agrees with Reiter's Default Logic; by contrast, Constrained Default Logic (Delgrande et al., 1994) is semimonotonic.

Semimonotonicity is regarded as a desirable property, because it is a form of locality: Extensions for large theories can be approximated by the formation of extensions for smaller ones. Reiter's Default Logic is semimonotonic in the case of normal theories, in which case we can apply the approximating process just mentioned. Although general extensions fail to give a semimonotonic framework for Default Logic, they still allow us to construct extensions "from below," although in a different sense. Moreover, in the case of seminormal theories, as we will see, the extension thus obtained is unique.

EXAMPLE 3.3.8 *(Weak orthogonality of extensions.) Consider the default theory with* $W = \emptyset$ *and* Δ *comprising the defaults*

$$\frac{: p \wedge \neg q}{\neg q}, \qquad \frac{: \neg p \wedge q}{\neg p}.$$

Orthogonality is the property that any two distinct extensions are inconsistent. Delgrande et al. (1994) proposed a similar notion, more appropriate in the context of Constrained Default Logic, i.e., weak orthogonality: This is the property according to which any two sets of constraints corresponding to distinct extensions are inconsistent.

Although weak orthogonality seems an appropriate feature in the case of the family of logics inspired by "commitment to the assumptions" (such as Delgrande et al., 1994; Lukaszewicz, 1988; Brewka, 1991), it appears to be less so in the present context. Indeed, it is possible to argue that (weak) orthogonality, or even maximality of extensions, is not a desirable feature given the three-valued intuition at the basis of the notion of general extension. In fact, orthogonality fails for general extensions, for the preceding theory has two maximal (mutually inconsistent) extensions, whose intersection is, however, again an extension.

EXAMPLE 3.3.9 *(Commitment to assumptions.) This is the "broken arms" example of Poole (1988). Consider the theory* (W, Δ) *in which* $W = \{\neg q \vee \neg s\}$ *and* Δ *comprises the defaults*

$$\frac{: p \wedge q}{p}, \qquad \frac{: r \wedge s}{r}.$$

Here the intuitive interpretation is as follows: We are told that either the left arm is broken or the right arm is broken $(\neg q \vee \neg s)$*; the first default asserts that a person's left arm is usable* (p) *unless it is broken* $(\neg q)$*, whereas the second default asserts that a person's right arm is usable* (r) *unless it is broken* $(\neg s)$.

Reiter's notion of extension here gives that both arms are usable, although we are told that at least one is broken. There is in fact a unique extension in Reiter's sense in which both defaults are fired, the justification of each one of them being consistent both with what is known and with the conclusions of fired defaults. The situation is essentially the same with the notion of general extension.

As Delgrande et al. (1994) point out, the problem here is that the consistency of each justification is tested individually, and not – as it were – wholesale. Indeed, this provides motivation for the adoption of the second construal of prerequisites that was mentioned, i.e., as working hypotheses rather than mere consistency conditions. However, if we do adopt the first (an equally feasible option, as Delgrande et al., 1994, admit), then this example loses much of its intuitive punch.

3.4 GROUNDED EXTENSIONS

In this section it is shown how to generalize the results of the previous section to default theories that might not be categorical. The treatment in this section reveals that prerequisites play an unexpectedly subtle role, in more than one way. It also turns out that the definitions and theorems in this section are somewhat more complicated than their analogs relative to categorical default theories. Accordingly, to simplify matters somewhat, with no loss in generality, we restrict our attention to default theories (W, Δ) for which W is propositionally consistent.

DEFINITION 3.4.1 *Let* (W, Δ) *be a default theory and* $\Gamma \subseteq \Delta$. *Then* $\mathsf{Ad}(\Gamma)$ *is the set of defaults (from* Δ*) that are* admissible *in* Γ*:*

$$\mathsf{Ad}(\Gamma) = \{\delta : C(\Gamma) \models_W P(\delta)\}.$$

We are going to be particularly interested in sets Γ of defaults such that $\Gamma \subseteq \mathsf{Ad}(\Gamma)$, i.e., every $\delta \in \Gamma$ is admissible in Γ. Recall that classical extensions for arbitrary default theories are required to be grounded, so that no extra assumptions are introduced that are not justified on the basis of what is known and the firing of default rules. Let us take a closer look at groundedness.

DEFINITION 3.4.2 *A set* Θ *of defaults is* grounded *if and only if* $\Theta = \bigcup_{n \geq 0} \Theta_n$*, where* $\Theta_0 = \emptyset$*, and*

$$\Theta_{n+1} = \{\delta \in \Theta : C(\Theta_n) \models_W P(\delta)\},$$

i.e., Θ_{n+1} *is the set of all defaults from* Θ *that are admissible in* Θ_n.

It is clear that if Γ is grounded then $\Gamma \subseteq \mathsf{Ad}(\Gamma)$. Notice that for any default theory (W, Δ) there will be several grounded sets $\Theta \subseteq \Delta$. But just because Θ is grounded does not mean that it is a set of defaults that we need to accept (that are triggered in some extension): Groundedness provides only a (weak) necessary condition; classical extensions are grounded in the preceding sense (proof in Section 3.6).

THEOREM 3.4.1 *Let* Γ *be a classical extension for a default theory* (W, Δ)*; then* Γ *is grounded.*

We are now ready to give the definition of general extensions for arbitrary default theories. We identify such extensions with triples $(\Gamma^+, \Gamma^-, \Gamma^*)$ of sets of defaults, in which Γ^+ is the set of defaults explicitly triggered, Γ^- is the set of defaults explicitly conflicted in Γ^+ or preempted in Γ^+, and Γ^* (the set of "potentially admissible" defaults) expresses the *degree of caution* of the extension. In general, the larger the Γ^*, the smaller the Γ^+, so that the size of Γ^* is proportional to the *caution* of Γ^+ and hence it is inversely related to its credulousness. In the formal development, as we will see (Theorem 3.4.4), this is reflected in the fact that general extensions can be obtained by an iterative process that is *monotonic* in Γ^+ and Γ^-, but *antimonotonic* in Γ^*.

DEFINITION 3.4.3 A general extension *for a default theory* (W, Δ) *is a triple* $(\Gamma^+, \Gamma^-, \Gamma^*)$ *of sets of defaults from* Δ, *such that*

- Γ^+ *and* Γ^- *are disjoint;*
- *the following two fixpoint equations are simultaneously satisfied:*

$$\begin{aligned}\Gamma^+ &= \{\delta : C(\Gamma^+) \models_W P(\delta) \;\&\; C(\Gamma^+) \not\models_W \neg C(\delta) \text{ and} \\ &\qquad C(\Gamma^* - \Gamma^-) \not\models_W \neg J(\delta)\}; \\ \Gamma^- &= \{\delta : C(\Gamma^+) \models_W \neg C(\delta) \text{ or } C(\Gamma^+) \models_W \neg J(\delta)\};\end{aligned}$$

- $\mathsf{Ad}(\Gamma^+) \subseteq \Gamma^* \subseteq \{\delta : C(\Gamma^+) \not\models_W \neg P(\delta)\}$.

In words: Γ^+ *is the set of all defaults admissible in* Γ^+ *but neither conflicted in* Γ^+ *nor preempted in* $(\Gamma^* - \Gamma^-)$; Γ^- *is the set of all defaults either conflicted or preempted in* Γ^+; *and* Γ^* *is a set of defaults containing all defaults admissible in* Γ^+ *and only defaults whose prerequisites are consistent with* $C(\Gamma^+)$ *(the latter condition is intended to capture the fact that* δ *is "potentially" admissible in* Γ^+*).*

From the definition it follows immediately that if $(\Gamma^+, \Gamma^-, \Gamma^*)$ is a general extension, every default δ in Γ^+ is admissible in Γ^+, i.e., $\Gamma^+ \subseteq \mathsf{Ad}(\Gamma^+)$. Recall that we restricted out attention to default theories (W, Δ) with W propositionally consistent: We can see now that nothing is lost by that assumption, for if W is propositionally inconsistent, then $(\emptyset, \Delta, \emptyset)$ is the only extension of the theory. The following theorem shows that this definition of general extension coincides with Definition 3.2.3 in the case of categorical theories.

THEOREM 3.4.2 *Let* (W, Δ) *be a categorical default theory and let* (Γ^+, Γ^-) *be an extension in the sense of Definition 3.2.3. Then* $(\Gamma^+, \Gamma^-, \Delta)$ *is an extension in the sense of Definition 3.4.3.*

We are going to be interested in extensions that are *minimal* according to the ordering subsequently defined.

DEFINITION 3.4.4 *Let* $(\Gamma^+, \Gamma^-, \Gamma^*)$ *and* $(\Theta^+, \Theta^-, \Theta^*)$ *be 3-tuples of sets of defaults. Define an ordering by putting* $(\Gamma^+, \Gamma^-, \Gamma^*) \leq (\Theta^+, \Theta^-, \Theta^*)$ *iff*

$$\Gamma^+ \subseteq \Theta^+, \quad \Gamma^- \subseteq \Theta^-, \quad \Theta^* \subseteq \Gamma^*.$$

We now state the analogs of Theorems 3.2.2 and 3.2.3 (proofs in Section 3.6).

THEOREM 3.4.3 *Let* $\Gamma^+ = \bigcup_{n \geq 0} \Gamma_n^+$ *be a classical extension for a default theory* (W, Δ). *Put*

$$\Gamma^- = \{\delta : \delta \text{ conflicted or preempted in } \Gamma^+\};$$
$$\Gamma^* = \{\delta : \delta \text{ admissible in } \Gamma^+\} = \mathsf{Ad}(\Gamma^+).$$

Then $(\Gamma^+, \Gamma^-, \Gamma^*)$ *is a general extension for* (W, Δ).

THEOREM 3.4.4 *Let* (W, Δ) *be a default theory. Then*

1. (W, Δ) *has an iteratively definable general extension,*
2. *such an extension is* $\leq$ *minimal, and*
3. *any minimal extension for* (W, Δ) *is grounded.*

THEOREM 3.4.5 *Let* $(\Gamma^+, \Gamma^-, \Gamma^*)$ *be a* $\leq$*-minimal extension for a default theory* (W, Δ). *Then* $(\Gamma^+, \Gamma^-, \Gamma^*)$ *can be obtained as the limit of an inductive construction of the kind given in the proof of Theorem 3.4.4.*

3.5 EXAMPLES, CONTINUED

EXAMPLE 3.5.1 *(Cumulativity.) Consider the theory* (W, Δ), *in which* $W = \emptyset$ *and* Δ *comprises the two defaults*

$$\delta_1 = \frac{: p}{p}, \qquad \delta_2 = \frac{p \vee q : \neg p}{\neg p}.$$

Cumulativity is the property that if a default is triggered in all the extensions of a theory, adding that default's conclusion to the world description W should give a theory that has the same extensions as the original one. This is related to the property of cautious monotonicity (if a "theorem" is added back to the set of facts from which it was deduced, then the set of "theorems" should not change), but it is not equivalent to it, as we will see.

Using Reiter's notion of extension in the preceding example (as we have seen in Section 1.3), cumulativity fails: The preceding theory has one extension Γ triggering the first default, and therefore $C(\Gamma) \models_W p \vee q$; but the second default cannot be triggered because it is preempted. Now consider the theory $(W \cup \{p \vee q\}, \Delta)$. This modified theory gains an extension, in which the second default rather than the first is triggered.

Let us see what happens with general extensions. The theory (W, Δ) has two general extensions: a minimal one $(\emptyset, \emptyset, \Delta)$ and a nonminimal one $(\delta_1, \delta_2, \delta_2)$, in which δ_1 is triggered and δ_2 ruled out.

When we add $p \vee q$ back into the world description W, we obtain a new extension, $(\delta_2, \delta_1, \delta_1)$, triggering the second default and ruling out the first. Therefore cumulativity fails also for the notion of general extension. Notice that this counterexample to cumulativity is also exactly the counterexample used by Makinson (1994) to establish that defeasible consequence defined as the intersection of classical extensions is not cautiously monotonic. However, as we will see in Theorem 4.1.6, this does not preclude Cautious Monotonicity from holding when defeasible consequence is defined as the intersection of general extensions, as adding the "theorem" $p \vee q$ back into the world description W gives *new* extensions, but no new *minimal* ones. In particular, although the new theory has more extensions than the old one, the *intersection* of the extensions (i.e., the set defaults triggered in *every* extension) is the same.

It is worth noting here that Brewka's Cumulative Default Logic is, of course, cumulative, whereas for the Constrained Default Logic of Delgrande et al. cumulativity fails except in the categorical (prerequisite-free) case.

EXAMPLE 3.5.2 *Consider a default theory* (W, Δ), *in which* W *is empty and* Δ *comprises the two defaults*

$$\delta_1 = \frac{: p}{p}, \qquad \delta_2 = \frac{q : \neg p}{\neg p}.$$

This example is particularly interesting, as it highlights the intuitions behind the present approach and the extent to which it delivers counterintuitive results. The point is that in the (unique) minimal extension of (W, Δ), neither default is triggered and neither is ruled out. This is due to the basic "skeptical" intuition behind general extensions: Defaults that are *potentially* preempted are not triggered. In the minimal extension, δ_1

is potentially preempted by $\neg p$ (the conclusion of δ_2), and δ_2 is potentially preempted by p (the conclusion of δ_1). The way things are set up, it does not matter that there is no way δ_2 will ever be triggered, as its prerequisite q is not entailed by the conclusions of any of the defaults in Δ.

To some extent this state of affairs is unsatisfying: Because δ_2 cannot possibly be triggered, it should not be allowed to prevent the triggering of δ_1. Let us notice that there seem to be different intuitions at work here: On the one hand there is the idea that "potentially preempted" defaults should not be triggered; on the other hand there is the intuition that defaults that are not even potentially admissible should be ruled out. In some cases, such as the present one, these intuitions appear to be in conflict.

Notice that the theory also has a *nonminimal* extension that avoids the problem by triggering δ_1 and (consequently) ruling out δ_2. In such an extension, we would have $\Gamma^+ = \{\delta_1\}$, $\Gamma^- = \{\delta_2\}$, and $\Gamma^* = \{\delta_1\}$. Observe also that there is no extension $(\Theta^+, \Theta^-, \Theta^*)$ with $\delta_2 \in \Theta^+$, because δ_2 would not be admissible in such a Θ^+.

Example 3.5.2 seems to be more far reaching than this. Consider the same theory with δ_1 and δ_2 as before, but now put $W = \{q\}$. Here is an interpretation in terms of a block world: Suppose $p = \mathsf{onTable}$ and $q = \mathsf{HeavyBlock}$. Then the theory appears to say that blocks are, by default, on the table, but heavy blocks are not. Because W tells us that there is a heavy block, the conclusion we would like to draw is that the heavy block is not on the table ($\neg p$). Information about heavy blocks is more specific than information about blocks. But under the present interpretation the theory has three general extensions, the one in which no default is triggered and two more, each triggering one of the defaults.

Of course, neither the present approach nor Reiter's is in a position to distinguish semantically between categorical defaults and defaults whose prerequisite is directly entailed by W. There is then no way to privilege the extension triggering δ_2 over the one triggering δ_1, which would be required to properly account for the specificity intuition. Against this, though, we should note that nothing in the theory tells us that information about heavy blocks is more specific than information about blocks; this is just a by-product of the particular interpretation we have chosen. Still, the fact that an extension triggering δ_2 might be desirable on some interpretations of the theory is another indicator of the interest of nonminimal extensions. For the time being, this is where we are going to leave the question of nonminimal extensions: We will come back to this question in Section 4.2, and indicate alternative ways in which this problem can be solved.

EXAMPLE 3.5.3 *Consider the default theory* (W, Δ) *with* $W = \{p\}$ *and* Δ *comprising the following defaults:*

$$\delta_1 = \frac{p : q}{q},$$
$$\delta_2 = \frac{q : q}{r},$$
$$\delta_3 = \frac{q : q}{\neg r},$$
$$\delta_4 = \frac{: \neg q}{s},$$
$$\delta_5 = \frac{: \neg s}{t}.$$

This example is intended to highlight features of general extensions that are not brought out by the previous examples. Let us consider the different defaults. Default δ_1 is always triggered: Its prerequisite is entailed by W, and it cannot be preempted because the negation of its justification q is not entailed by the conclusions of any of the defaults. Things are different for the pair of defaults δ_2 and δ_3: Once δ_1 is triggered, they are both admissible, and neither is (potentially) preempted. However, they cannot both be triggered because they have contradictory conclusions. This state of affairs is a "diamond," analogous to the one of Fig. 1.2 (Section 1.4). At the diamond, extensions split. Finally consider the two defaults δ_4 and δ_5: The latter cannot be triggered initially, because it was potentially preempted by δ_4. However, as soon as δ_1 is triggered, δ_4 is ruled out (because preempted), removing any obstacles to the triggering of δ_5. In conclusion, the theory (W, Δ) has two minimal extensions, one extension triggering δ_1, δ_2, and δ_5; and the other extension triggering δ_1, δ_3, and δ_5. In the first extension δ_2 and δ_4 are ruled out and in the second δ_3 and δ_4 are ruled out.

3.6 PROOFS OF SELECTED THEOREMS

THEOREM 3.2.2 Let Γ^+ be a classical extension for a categorical default theory (W, Δ), and put

$$\Gamma^- = \{\delta : \delta \text{ conflicted or preempted in } \Gamma^+\};$$

then (Γ^+, Γ^-) is a general extension for (W, Δ).

Proof. From the hypothesis that Γ^+ is a classical extension we know that it satisfies

$$\Gamma^+ = \{\delta \in \Delta : C(\Gamma^+) \not\models_W \neg J(\delta)\}.$$

First we note that Γ^+ and Γ^- are disjoint: If δ belongs to both Γ^+ and Γ^- then (from the latter) we have that either δ is conflicted in Γ^+ or δ is preempted in Γ^+. The second alternative is impossible, as δ is in Γ^+ (and hence *not* preempted in Γ^+); therefore it can only be that δ is conflicted in Γ^+, i.e., $C(\Gamma^+) \models_W \neg C(\delta)$. Because δ itself is in Γ^+, we have that $C(\Gamma^+)$ is W-inconsistent and any default is preempted in an inconsistent set of sentences. It would follow that δ is preempted in Γ^+ after all, which we have already ruled out. We conclude that there cannot be any such δ.

Therefore we need to show that the pair (Γ^+, Γ^-) satisfies the pair of fixpoint equations of Definition 3.2.3. In turn, we observe that the second of such equations, namely

$$\Gamma^- = \{\delta : C(\Gamma^+) \models_W \neg C(\delta) \text{ or } C(\Gamma^+) \models_W \neg J(\delta)\},$$

is satisfied by definition of Γ^-; so we need to establish that

$$\Gamma^+ = \{\delta : C(\Gamma^+) \not\models_W \neg C(\delta) \,\&\, C(\Delta - \Gamma^-) \not\models_W \neg J(\delta)\}.$$

What we need to show is that δ is not preempted in Γ^+ if and only if it is neither conflicted in Γ^+ nor preempted in $\Delta - \Gamma^-$:

$$C(\Gamma^+) \not\models_W \neg J(\delta) \iff C(\Gamma^+) \not\models_W \neg C(\delta) \,\&\, C(\Delta - \Gamma^-) \not\models_W \neg J(\delta).$$

In turn, this claim breaks down into the following parts, which need to be separately established:

1. $C(\Gamma^+) \models_W \neg J(\delta)$ only if $C(\Gamma^+) \models_W \neg C(\delta)$ or $C(\Delta - \Gamma^-) \models_W \neg J(\delta)$;
2. $C(\Gamma^+) \models_W \neg C(\delta)$ only if $C(\Gamma^+) \models_W \neg J(\delta)$;
3. $C(\Delta - \Gamma^-) \models_W \neg J(\delta)$ only if $C(\Gamma^+) \models_W \neg J(\delta)$.

Part 1 establishes the converse implication, whereas parts 2 and 3 together suffice for the direct implication.

For part 1, assume $C(\Gamma^+) \models_W \neg J(\delta)$; then there are defaults $\delta_1, \ldots, \delta_n \in \Gamma^+$ such that $C(\delta_1), \ldots, C(\delta_n) \models_W \neg J(\delta)$; because Γ^+ is a classical extension, $\delta_1, \ldots, \delta_n$ are neither conflicted nor preempted in Γ^+, whence by definition of Γ^- we have $\delta_1, \ldots, \delta_n \notin \Gamma^-$, which is to say $C(\Delta - \Gamma^-) \models_W \neg J(\delta)$. Part 1 follows.

For part 2, suppose $C(\Gamma^+) \models_W \neg C(\delta)$. We can assume that $C(\Gamma^+)$ is consistent, because otherwise the conclusion $C(\Gamma^+) \models_W \neg J(\delta)$ follows immediately. Then, if $C(\Gamma^+)$ is consistent, we have $\delta \notin \Gamma^+$ [because $C(\Gamma^+) \models_W \neg C(\delta)$], and because Γ^+ is a classical extension, also $C(\Gamma^+) \models_W \neg J(\delta)$, which is the desired conclusion.

Finally, for part 3, suppose δ is preempted in $C(\Delta - \Gamma^-)$, i.e., $C(\Delta - \Gamma^-) \models_W \neg J(\delta)$. Then there are $\delta_1, \ldots, \delta_n \notin \Gamma^-$ such that $C(\delta_1), \ldots, C(\delta_n) \models_W \neg J(\delta)$. By the definition of Γ^-, we know that $\delta_1, \ldots, \delta_n$ are neither conflicted nor preempted in Γ^+, and because Γ^+ is a classical extension, we have $\delta_1, \ldots, \delta_n \in \Gamma^+$ by Theorem 3.2.1. The desired conclusion $C(\Gamma^+) \models_W \neg J(\delta)$ follows. ■

THEOREM 3.2.3 Every categorical default theory has a general extension.

Proof. Let (W, Δ) be a categorical default theory; we can assume that W is consistent, because otherwise $(\emptyset, \Delta)$ is an extension. We define a general extension (Γ^+, Γ^-), where Γ^+ is the union of a sequence $\Gamma_0^+, \Gamma_1^+, \ldots$, and Γ^- is the union of the sequence $\Gamma_0^-, \Gamma_1^-, \ldots$; in turn, the sets Γ_n^+ and Γ_n^- are inductively defined with an iterative process similar to the one in Horty (1994) or Antonelli (1997).

For the base case put $\Gamma_0^+ = \Gamma_0^- = \emptyset$, which are obviously disjoint. For the inductive step, let Γ_{n+1}^+ be a *maximal* set of defaults such that

1. $C(\Gamma_n^+ \cup \Gamma_{n+1}^+)$ is consistent,
2. no default $\delta \in \Gamma_{n+1}^+$ is preempted in $\Delta - \Gamma_n^-$.

(Observe that such a maximal set of defaults always exists, although it might not be unique.) For Γ_{n+1}^-, put

$$\Gamma_{n+1}^- = \{\delta : \delta \text{ preempted or conflicted in } \Gamma_{n+1}^+\}.$$

As in the proof of Theorem 3.2.2, it follows that Γ_{n+1}^+ and Γ_{n+1}^- are disjoint.

Next, we show that $\Gamma_n^+ \subseteq \Gamma_{n+1}^+$ and $\Gamma_n^- \subseteq \Gamma_{n+1}^-$, by induction on n. The base case for $n = 0$ is trivial. For the inductive step, we assume that $\Gamma_n^+ \subseteq \Gamma_{n+1}^+$ and $\Gamma_n^- \subseteq \Gamma_{n+1}^-$ in order to show that $\Gamma_{n+1}^+ \subseteq \Gamma_{n+2}^+$ and $\Gamma_{n+1}^- \subseteq \Gamma_{n+2}^-$.

Ad $\Gamma_{n+1}^+ \subseteq \Gamma_{n+2}^+$. Let $\delta \in \Gamma_{n+1}^+$. First we show that δ is not preempted in $\Delta - \Gamma_{n+1}^-$, for if δ were so preempted, then $C(\Delta - \Gamma_{n+1}^-) \models_W \neg J(\delta)$, and because $(\Delta - \Gamma_{n+1}^-) \subseteq (\Delta - \Gamma_n^-)$ by the inductive hypothesis, also $C(\Delta - \Gamma_n^-) \models_W \neg J(\delta)$, which is impossible if $\delta \in \Gamma_{n+1}^+$. Moreover, by the consistency of $C(\Gamma_{n+1}^+ \cup \Gamma_{n+2}^+)$, we have that δ cannot be inconsistent with $C(\Gamma_{n+1}^+ \cup \Gamma_{n+2}^+)$. By the maximality of Γ_{n+2}^+, it follows that $\delta \in \Gamma_{n+2}^+$.

Ad $\Gamma^-_{n+1} \subseteq \Gamma^-_{n+2}$. Suppose $\delta \in \Gamma^-_{n+1}$. Then δ is either conflicted or preempted in Γ^+_{n+1}, and because $\Gamma^+_{n+1} \subseteq \Gamma^+_{n+2}$, we have that δ is also conflicted or, respectively, preempted in Γ^+_{n+2}, so that $\delta \in \Gamma^-_{n+2}$.

This shows that the sequences Γ^+_n and Γ^-_n (for $n \geq 0$) are increasing. Put

$$\Gamma^+ = \bigcup_{n \geq 0} \Gamma^+_n,$$

$$\Gamma^- = \bigcup_{n \geq 0} \Gamma^-_n.$$

We verify that (Γ^+, Γ^-) is an extension for (W, Δ). First we observe that Γ^+ and Γ^- are disjoint (any common member δ would have had to be put in at some stage n, but this is impossible). Next, we check that the following equation is satisfied:

$$\Gamma^+ = \{\delta : C(\Gamma^+) \not\models_W \neg C(\delta) \text{ and } C(\Delta - \Gamma^-) \not\models_W \neg J(\delta)\}.$$

In one direction, let $\delta \in \Gamma^+$ and choose k such that $\delta \in \Gamma^+_k$. First we show that δ cannot be conflicted in Γ^+, for otherwise it would have been conflicted in some Γ^+_n, and in particular also in $\Gamma^+_{\max(n,k)}$, which would then have been inconsistent, against construction. Moreover, δ cannot be preempted in $\Delta - \Gamma^-$: If it were so preempted, then for some $\delta_1, \ldots, \delta_m \notin \Gamma^-$ we would have $C(\delta_1), \ldots, C(\delta_m) \models_W \neg J(\delta)$. Because the sequence Γ^-_n is increasing, the sequence $(\Delta - \Gamma^-_n)$ is *decreasing*, so that δ would have been preempted also in some (indeed, every) $\Delta - \Gamma^-_n$.

For the other direction, we need to show that if δ is neither conflicted in Γ^+ nor preempted in $\Delta - \Gamma^-$ then $\delta \in \Gamma^+$. First observe that if δ is not conflicted in Γ^+, then it cannot be conflicted in any Γ^+_n. So if $\delta \notin \Gamma^+$ it must be because δ is preempted in *every* $\Delta - \Gamma^-_n$. So for every $n \geq 0$, there are $\delta_1, \ldots, \delta_m \notin \Gamma^-_n$ such that $C(\delta_1), \ldots, C(\delta_m) \models_W \neg J(\delta)$. *Now for the first time we use the finiteness hypothesis for* Δ: Because Δ is finite, there are finitely many m-tuples of defaults. Because the sequence $(\Delta - \Gamma^-_n)$ is decreasing, some m-tuple of defaults preempting δ must be in every $(\Delta - \Gamma^-_n)$, and hence also in $(\Delta - \Gamma^-)$, against hypothesis. So there must be some k such that every m-tuple $\delta_1, \ldots, \delta_m$ of defaults preempting δ has a member in Γ^-_k. Then δ is not preempted in $(\Delta - \Gamma^-_k)$, whence $\delta \in \Gamma^+_{k+1} \subseteq \Gamma^+$, as required.

For the second equation, we need to verify that

$$\Gamma^- = \{\delta : C(\Gamma^+) \models_W \neg C(\delta) \text{ or } C(\Gamma^+) \models_W \neg J(\delta)\}.$$

In one direction, suppose $\delta \in \Gamma^-$; then for some n, $\delta \in \Gamma^-_{n+1}$, so that δ is preempted or conflicted in Γ^+_{n+1}. This implies that δ is also preempted or conflicted in Γ^+. For the other direction, suppose δ is preempted in Γ^+. So there are defaults $\delta_1, \ldots, \delta_m \in \Gamma^+$ such that $C(\delta_1), \ldots, C(\delta_m) \models_W \neg J(\delta)$. Find n large enough such that $\delta_1, \ldots, \delta_m \in \Gamma_n^+$. Then $\delta \in \Gamma_n^- \subseteq \Gamma^-$. The case for δ conflicted is similar. ∎

THEOREM 3.4.1 Let Γ be a classical extension for a default theory (W, Δ); then Γ is grounded.

Proof. Because Γ is classical, $\Gamma = \bigcup_{n\geq 0} \Gamma_n$, where $\Gamma_0 = \emptyset$ and

$$\Gamma_{n+1} = \{\delta : \delta \text{ admissible in } \Gamma_n \text{ and not preempted in } \Gamma\}.$$

To show that Γ is grounded, put $\Theta_0 = \emptyset$ and $\Theta_{n+1} = \{\delta \in \Gamma : \delta \text{ admissible in } \Theta_n\}$.

We need to show that $\Gamma = \bigcup_{n\geq 0} \Theta_n$, i.e.,

$$\bigcup_{n\geq 0} \Gamma_n = \bigcup_{n\geq 0} \Theta_n.$$

First we observe that the inclusion $\Theta_n \subseteq \Gamma$ holds for every n, by definition of Θ_n. It follows that $\bigcup_{n\geq 0} \Theta_n \subseteq \Gamma$.

Second, we show that $\Gamma \subseteq \bigcup_{n\geq 0} \Theta_n$. In turn, it suffices to show that $\Gamma_p \subseteq \bigcup_{n\geq 0} \Theta_n$ for every p. We proceed by induction on p. For the base case, $\Gamma_0 = \emptyset \subseteq \bigcup_{n\geq 0} \Theta_n$. For the inductive step we assume that $\Gamma_p \subseteq \bigcup_{n\geq 0} \Theta_n$. Suppose $\delta \in \Gamma_{p+1}$: In particular, δ is admissible in Γ_p. So there are defaults $\delta_1, \ldots, \delta_m \in \Gamma_p$ such that $C(\delta_1), \ldots, C(\delta_m) \models_W P(\delta)$. By the inductive hypothesis $\Gamma_p \subseteq \bigcup_{n\geq 0} \Theta_n$, so there must be $q \geq 0$ such that $\delta_1, \ldots, \delta_m \in \Theta_q$. Then $\delta \in \Theta_{q+1} \subseteq \bigcup_{n\geq 0} \Theta_n$, as required.

If we now put $\Theta = \bigcup_{n\geq 0} \Theta_n$, we have $\Gamma = \Theta$. Then we have $\Theta_0 = \emptyset$ and $\Theta_{n+1} = \{\delta \in \Theta : \delta \text{ admissible in } \Theta_n\}$. So Θ is grounded and hence so is Γ. ∎

THEOREM 3.4.3 Let $\Gamma^+ = \bigcup_{n\geq 0} \Gamma_n^+$ be a classical extension for a default theory (W, Δ). Put

$$\Gamma^- = \{\delta : \delta \text{ conflicted or preempted in } \Gamma^+\};$$
$$\Gamma^* = \{\delta : \delta \text{ admissible in } \Gamma^+\} = \mathsf{Ad}(\Gamma^+).$$

Then $(\Gamma^+, \Gamma^-, \Gamma^*)$ is a general extension for (W, Δ).

Proof. We modify the proof of Theorem 3.2.2 as needed. From the hypothesis that Γ^+ is a classical extension it follows that if $\delta \in \Gamma^-$ then

$\delta \notin \Gamma^+$, i.e., Γ^+ and Γ^- are disjoint. Moreover, we know that $\Gamma^+ = \bigcup_{n\geq 0} \Gamma_n^+$, where $\Gamma_0^+ = \emptyset$, and

$$\Gamma_{n+1}^+ = \{\delta \in \Delta : C(\Gamma_n^+) \models_W P(\delta) \text{ and } C(\Gamma^+) \not\models_W \neg J(\delta)\}.$$

So we need to show that the triple $(\Gamma^+, \Gamma^-, \Gamma^*)$ satisfies the conditions of Definition 3.4.3. In turn, we observe that the condition

$$\Gamma^- = \{\delta : C(\Gamma^+) \models_W \neg C(\delta) \text{ or } C(\Gamma^+) \models_W \neg J(\delta)\}$$

is satisfied by definition of Γ^-, whereas the condition

$$\mathsf{Ad}(\Gamma^+) \subseteq \Gamma^* \subseteq \{\delta : C(\Gamma^+) \not\models_W \neg P(\delta)\}$$

derives from the hypothesis that W is consistent and hence $C(\Gamma^+)$ is W-consistent. So we need to establish that

$$\Gamma^+ = \{\delta : C(\Gamma^+) \models_W P(\delta) \;\&\; C(\Gamma^+) \not\models_W \neg C(\delta) \text{ and } \\ C(\Gamma^* - \Gamma^-) \not\models_W \neg J(\delta)\}.$$

In other words, what we need to show is that $\delta \in \Gamma^+$ if and only if it is admissible in Γ^+, but neither conflicted in Γ^+ nor preempted in $\Gamma^* - \Gamma^-$. In turn, this claim breaks down into the following four parts:

1. $\delta \notin \Gamma^+$ only if $C(\Gamma^+) \not\models_W P(\delta)$ or $C(\Gamma^+) \models_W \neg C(\delta)$ or $C(\Gamma^* - \Gamma^-) \models_W \neg J(\delta)$,
2. $C(\Gamma^+) \models_W \neg C(\delta)$ only if $\delta \notin \Gamma^+$,
3. $C(\Gamma^* - \Gamma^-) \models_W \neg J(\delta)$ only if $\delta \notin \Gamma^+$,
4. $C(\Gamma^+) \not\models_W P(\delta)$ only if $\delta \notin \Gamma^+$.

For part 1, assume that $\delta \notin \Gamma^+$. Because Γ^+ is classical, by Theorem 3.2.1, either (1) δ is not admissible in Γ^+ or (2) it is conflicted in Γ^+, or (3) it is preempted in Γ^+. The conclusion follows immediately in cases (1) and (2). For case (3) we need to show that if δ is preempted in Γ^+, then it is preempted in $C(\Gamma^* - \Gamma^-)$. So suppose δ is preempted in Γ^+: Then there are defaults $\delta_1, \ldots, \delta_n \in \Gamma^+$ such that $C(\delta_1), \ldots, C(\delta_n) \models_W \neg J(\delta)$. Because Γ^+ is a classical extension, $\delta_1, \ldots, \delta_n$ are neither conflicted nor preempted in Γ^+, whence by definition of Γ^- we have $\delta_1, \ldots, \delta_n \notin \Gamma^-$. Moreover, again because Γ^+ is classical, we have that $\delta_1, \ldots, \delta_n$ are all admissible in Γ^+, and hence $\delta_1, \ldots, \delta_n \in \Gamma^*$. It follows that $C(\Gamma^* - \Gamma^-) \models_W \neg J(\delta)$, whence part 1 follows.

For part 2, we have immediately that because Γ^+ is classical, the hypothesis that $\delta \in \Gamma^+$ implies that δ is not conflicted in Γ^+.

For part 3, suppose δ is preempted in $C(\Gamma^* - \Gamma^-)$, i.e.,

$$C(\Gamma^* - \Gamma^-) \models_W \neg J(\delta).$$

Then there are $\delta_1, \ldots, \delta_n \notin \Gamma^-$ such that $C(\delta_1), \ldots, C(\delta_n) \models_W \neg J(\delta)$. In particular, $\delta_1, \ldots, \delta_n$ belong to Γ^* and hence are admissible in Γ^+; moreover, by the definition of Γ^-, we know that $\delta_1, \ldots, \delta_n$ are neither conflicted nor preempted in Γ^+. So, because Γ^+ is a classical extension, we have $\delta_1, \ldots, \delta_n \in \Gamma^+$ by Theorem 3.2.1. The desired conclusion $C(\Gamma^+) \models_W \neg J(\delta)$ follows.

Finally, for part 4, if $\delta \in \Gamma^+$ then δ is admissible in Γ^+, as Γ^+ is a classical extension. ■

THEOREM 3.4.4 Let (W, Δ) be a default theory. Then

1. every default theory has an iteratively definable general extension,
2. such an extension is $\leq$ minimal, and
3. any minimal extension for (W, Δ) is grounded.

Proof. Part (1) is similar to the proof of Theorem 3.2.3. Let (W, Δ) be a default theory. We define a general extension $(\Gamma^+, \Gamma^-, \Gamma^*)$, where $\Gamma^+ = \bigcup_{n\geq 0} \Gamma_n^+$ and $\Gamma^- = \bigcup_{n\geq 0} \Gamma_n^-$, whereas $\Gamma^* = \bigcap_{n\geq 0} \Gamma_n^*$.

For the base case put $\Gamma_0^+ = \emptyset$, $\Gamma_0^- = \{\delta : \delta \text{ conflicted or preempted in } \Gamma_0^+\}$, and $\Gamma_0^* = \{\delta : \Gamma_0^+ \not\models_W \neg P(\delta)\}$.

For the inductive step, let Γ_{n+1}^+ be a *maximal* set of defaults such that

1. $C(\Gamma_n^+ \cup \Gamma_{n+1}^+)$ is consistent,
2. every $\delta \in \Gamma_{n+1}^+$ is admissible in Γ_n^+,
3. no default $\delta \in \Gamma_{n+1}^+$ is preempted in $\Gamma_n^* - \Gamma_n^-$.

(Observe again that such a maximal set of defaults need not be unique.) Moreover, put

$$\Gamma_{n+1}^- = \{\delta : \delta \text{ pre-empted or conflicted in } \Gamma_{n+1}^+\},$$
$$\Gamma_{n+1}^* = \{\delta : C(\Gamma_{n+1}^+) \not\models_W \neg P(\delta)\}.$$

Next, we show that the sequences Γ_n^+ and Γ_n^- are *increasing*, whereas the sequence Γ_n^* is *decreasing*. We proceed by induction on n and show that

1. $\Gamma_n^+ \subseteq \Gamma_{n+1}^+$,
2. $\Gamma_n^- \subseteq \Gamma_{n+1}^-$,
3. $\Gamma_{n+1}^* \subseteq \Gamma_n^*$.

Base case for $n = 0$: Trivially we have $\Gamma_0^+ \subseteq \Gamma_1^+$. Moreover, if $\delta \in \Gamma_0^-$, then δ is conflicted or preempted in Γ_0^+: Then it is also conflicted or preempted in Γ_1^+, so that $\delta \in \Gamma_1^-$. This shows that $\Gamma_0^- \subseteq \Gamma_1^-$. Finally, if $C(\Gamma_0^+) \models_W \neg P(\delta)$, then also $C(\Gamma_1^+) \models_W \neg P(\delta)$, so that $\Gamma_1^* \subseteq \Gamma_0^*$.

For the inductive step, assume $\Gamma_n^+ \subseteq \Gamma_{n+1}^+$ and $\Gamma_n^- \subseteq \Gamma_{n+1}^-$, as well as $\Gamma_{n+1}^* \subseteq \Gamma_n^*$, in order to show the analog properties for $n+1$ and $n+2$.

Ad $\Gamma_{n+1}^+ \subseteq \Gamma_{n+2}^+$. Let $\delta \in \Gamma_{n+1}^+$. First we show that δ is not preempted in $\Gamma_{n+1}^* - \Gamma_{n+1}^-$, for if δ were so preempted, then $C(\Gamma_{n+1}^* - \Gamma_{n+1}^-) \models_W \neg J(\delta)$, and because by the inductive hypothesis $(\Delta - \Gamma_{n+1}^-) \subseteq (\Delta - \Gamma_n^-)$, as well as $\Gamma_{n+1}^* \subseteq \Gamma_n^*$ also $C(\Gamma_n^* - \Gamma_n^-) \models_W \neg J(\delta)$, which is impossible if $\delta \in \Gamma_{n+1}^+$. Moreover, by the W-consistency of $C(\Gamma_{n+1}^+ \cup \Gamma_{n+2}^+)$, we have that δ cannot be W-inconsistent with $C(\Gamma_{n+1}^+ \cup \Gamma_{n+2}^+)$. By the maximality of Γ_{n+2}^+, it follows that $\delta \in \Gamma_{n+2}^+$.

Ad $\Gamma_{n+1}^- \subseteq \Gamma_{n+2}^-$. Suppose $\delta \in \Gamma_{n+1}^-$. Then δ is either conflicted or preempted in Γ_{n+1}^+, and because $\Gamma_{n+1}^+ \subseteq \Gamma_{n+2}^+$, we have that δ is also conflicted or, respectively, preempted in Γ_{n+2}^+, so that $\delta \in \Gamma_{n+2}^-$.

Finally, ad $\Gamma_{n+2}^* \subseteq \Gamma_{n+1}^*$. This follows from the fact that $\Gamma_{n+1}^+ \subseteq \Gamma_{n+2}^+$, so that if $C(\Gamma_{n+1}^+) \models_W \neg P(\delta)$ then also $C(\Gamma_{n+2}^+) \models_W \neg P(\delta)$.

This shows that the sequences Γ_n^+ and Γ_n^- are increasing whereas Γ_n^* is decreasing (for $n \geq 0$). It is not hard to see that the sequences Γ_n^+ and Γ_n^- are disjoint. Now put

$$\Gamma^+ = \bigcup_{n \geq 0} \Gamma_n^+,$$
$$\Gamma^- = \bigcup_{n \geq 0} \Gamma_n^-,$$
$$\Gamma^* = \bigcap_{n \geq 0} \Gamma_n^*.$$

Observe that $C(\Gamma^+)$ must be consistent, given the assumption that W is consistent. We verify that $(\Gamma^+, \Gamma^-, \Gamma^*)$ is an extension for (W, Δ). First we remark that the sets Γ^+ and Γ^- are disjoint, as they are the limits of disjoint sequences. Then we check that the following equation is satisfied:

$$\Gamma^+ = \{\delta : C(\Gamma^+) \models_W P(\delta) \;\&\; C(\Gamma^+) \not\models_W \neg C(\delta) \;\&\; C(\Gamma^* - \Gamma^-) \not\models_W \neg J(\delta)\}.$$

In one direction, let $\delta \in \Gamma^+$ and choose k such that $\delta \in \Gamma_k^+$. First we observe that δ cannot be conflicted in Γ^+, for otherwise it would have been conflicted in some Γ_n^+, and in particular also in $\Gamma_{\max(n,k)}^+$, which would then have been inconsistent, against construction. Moreover, δ must be admissible in Γ_k^+ and hence in Γ^+. Finally, δ cannot be preempted in $\Gamma^* - \Gamma^-$: If it were so preempted, then for some $\delta_1, \ldots, \delta_m \in (\Gamma^* - \Gamma^-)$ we would have $C(\delta_1), \ldots, C(\delta_m) \models_W \neg J(\delta)$. Because the sequence Γ_n^- is increasing, the sequence $(\Delta - \Gamma_n^-)$ is *decreasing*, so that δ would have been preempted also in every $(\Gamma_n^* - \Gamma_n^-)$, including $(\Gamma_k^* - \Gamma_k^-)$.

For the other direction, we need to show that if δ is admissible in Γ^+ but neither conflicted in Γ^+ nor preempted in $\Gamma^* - \Gamma^-$ then $\delta \in \Gamma^+$. First observe that if δ is not conflicted in Γ^+, then it cannot be conflicted in any Γ_n^+. Moreover, if δ is admissible in Γ^+ it must be admissible in Γ_k^+ for every k greater than or equal to some p. So if $\delta \notin \Gamma^+$ it must be because δ is preempted in *every* $\Gamma_{n+p}^* - \Gamma_{n+p}^-$. So for every $n \geq 0$, there are $\delta_1, \ldots, \delta_m \in (\Gamma_{n+p}^* - \Gamma_{n+p}^-)$ such that $C(\delta_1), \ldots, C(\delta_m) \models_W \neg J(\delta)$. *Now we use the finiteness hypothesis for* Δ*:* Because Δ is finite and the sequence $(\Gamma_n^* - \Gamma_n^-)$ is decreasing, there is some m-tuple $\delta_1, \ldots, \delta_m$ preempting δ and some $q \geq 0$ such that $\delta_1, \ldots, \delta_m \in (\Gamma_p^* - \Gamma_p^-)$ for all $p \geq q$. Then δ is preempted in $\Gamma^* - \Gamma^-$, against assumption. So $\delta \in \Gamma_{k+1}^+ \subseteq \Gamma^+$, as required.

For the second equation, we need to verify that

$$\Gamma^- = \{\delta : C(\Gamma^+) \models_W \neg C(\delta) \text{ or } C(\Gamma^+) \models_W \neg J(\delta)\}.$$

In one direction, suppose $\delta \in \Gamma^-$; then for some n, $\delta \in \Gamma_{n+1}^-$, so that δ is preempted or conflicted in Γ_{n+1}^+. This implies that δ is also preempted or conflicted in Γ^+. For the other direction, suppose δ is preempted in Γ^+. So there are defaults $\delta_1, \ldots, \delta_m \in \Gamma^+$ such that $C(\delta_1), \ldots, C(\delta_m) \models_W \neg J(\delta)$. Find n large enough such that $\delta_1, \ldots, \delta_m \in \Gamma_n^+$; then $\delta \in \Gamma_n^- \subseteq \Gamma^-$. The case for δ conflicted is similar.

Finally, the condition

$$\mathsf{Ad}(\Gamma^+) \subseteq \Gamma^* \subseteq \{\delta : C(\Gamma^+) \not\models_W \neg P(\delta)\}$$

is satisfied by construction. For the first inclusion we observe that if δ is admissible in Γ^+ then $C(\Gamma^+) \not\models_W \neg P(\delta)$, given that $C(\Gamma^+)$ is consistent. For the second inclusion, if $C(\Gamma^+) \models_W \neg P(\delta)$, then for some n, $C(\Gamma_n^+) \not\models_W \neg P(\delta)$, whence $\delta \notin \Gamma_n^*$; this implies that $\delta \notin \bigcap_{n \geq 0} \Gamma_n^* = \Gamma^*$.

We now take up part (2): We need to show that the extension obtained in the construction previously given is $\leq$ minimal. We do this by assuming that, for some extension $(\Theta^+, \Theta^-, \Theta^*)$, we have

$$(\Theta^+, \Theta^-, \Theta^*) \leq (\Gamma^+, \Gamma^-, \Gamma^*),$$

and by showing that $(\Theta^+, \Theta^-, \Theta^*) = (\Gamma^+, \Gamma^-, \Gamma^*)$. In turn, by induction on n, it suffices to show that

1. $\Gamma_n^+ \subseteq \Theta^+$,
2. $\Gamma_n^- \subseteq \Theta^-$,
3. $\Theta^* \subseteq \Gamma_n^*$.

In the case for $n = 0$, parts 1 and 2 are immediate. To show $\Theta^* \subseteq \Gamma_0^*$ suppose that $\delta \in \Theta^*$. Now, Γ_0^* is the set of defaults whose prerequisite is W-consistent (with Γ_0^+). If $P(\delta)$ were W-inconsistent, then $C(\Theta^+) \models_W \neg P(\delta)$, which implies that $\delta \notin \Theta^*$.

For the inductive step, assume parts 1–3 for n in order to establish them for $n+1$. From parts 2 and 3 for n we have $(\Theta^* - \Theta^-) \subseteq (\Gamma_n^* - \Gamma_n^-)$. Now suppose $\delta \in \Gamma_{n+1}^+$; then (a) δ is admissible in Γ_n^+ and hence also in Θ^+, (b) δ is not preempted in $(\Gamma_n^* - \Gamma_n^-)$ and hence not preempted in $(\Theta_n^* - \Theta_n^-)$, either. It remains to show that δ is not conflicted in Θ^+; if it were so conflicted, since by hypothesis $\Theta^+ \subseteq \Gamma^+$, δ would also be conflicted in Γ_m^+, for some $m \geq n+1$. Because $\delta \in \Gamma_{n+1}^+ \subseteq \Gamma_m^+$, we would have that $C(\Gamma_m^+)$ is W-inconsistent, which is impossible in our construction. This concludes the inductive step for part 1.

The inductive step for part 2 follows immediately from the inductive hypothesis. For part 3, we need to show that $\Theta^* \subseteq \Gamma_{n+1}^*$. If $\delta \notin \Gamma_{n+1}^*$, then $C(\Gamma_{n+1}^+) \models_W \neg P(\delta)$, and because $\Gamma_{n+1}^+ \subseteq \Theta^+$, also $C(\Theta^+) \models_W \neg P(\delta)$, so that $\delta \notin \Theta^*$.

Finally, for part 3 of the theorem, we show that Γ^+ is grounded. Put $\Theta_0 = \emptyset$, and $\Theta_{n+1} = \{\delta \in \Gamma^+ : \delta \text{ admissible in } \Theta_n\}$. If we let $\Theta = \bigcup_{n \geq 0} \Theta_n$, then, as before, it suffices to show that $\Theta = \Gamma^+$. As in Theorem 3.4.1, $\Theta \subseteq \Gamma^+$ is immediate. For the converse inclusion, we can show by induction on n that $\Gamma_n^+ \subseteq \Theta$, whence $\bigcup_{n \geq 0} \Gamma_n^+ \subseteq \Theta$. ■

THEOREM 3.4.5 Let $(\Gamma^+, \Gamma^-, \Gamma^*)$ be a $\leq$-minimal extension for a default theory (W, Δ). Then $(\Gamma^+, \Gamma^-, \Gamma^*)$ can be obtained as the limit of an inductive construction of the kind given in the proof of Theorem 3.4.4.

Proof. Suppose we have defined sets of defaults Θ_n^+, Θ_n^-, and Θ_n^* (the first two subsets of Γ^+, Γ^- respectively, and the third a superset of Γ^*), in such a way as to conform to the construction of Theorem 3.4.4. Then, if we put

$$\Theta^+ = \bigcup_{n \geq 0} \Theta_n^+, \quad \Theta^- = \bigcup_{n \geq 0} \Theta_n^*, \quad \Theta^* = \bigcap_{n \geq 0} \Theta_n^*,$$

we have that $(\Theta^+, \Theta^-, \Theta^*) \leq (\Gamma^+, \Gamma^-, \Gamma^*)$, and moreover $(\Theta^+, \Theta^-, \Theta^*)$ is an extension. By minimality of $(\Gamma^+, \Gamma^-, \Gamma^*)$, we have

$$(\Theta^+, \Theta^-, \Theta^*) = (\Gamma^+, \Gamma^-, \Gamma^*),$$

and the conclusion follows.

So we need to show how to define Θ_n^+, Θ_n^-, and Θ_n^*. We define $\Theta_0^+ = \Theta_0^- = \emptyset$, and $\Theta_0^* =$ the set of defaults whose prerequisite is consistent. For the inductive step, we pick as Θ_{n+1}^+ the set of defaults, *all drawn from* Γ^+, that are admissible at the previous stage:

$$\Theta_{n+1}^+ = \{\delta \in \Gamma^+ : C(\Theta_n^+) \models_W P(\delta)\}.$$

To show that this conforms to the construction, we need to establish that Θ_{n+1}^+ is a maximal set of defaults having the following properties: (1) $C(\Theta_n^+ \cup \Theta_{n+1}^+)$ is W-consistent; (2) every $\delta \in \Theta_{n+1}^+$ is admissible in Θ_n^+; and (3) no default in Θ_{n+1}^+ is preempted in $(\Theta_n^* - \Theta_n^-)$.

Clearly Θ_{n+1}^+ has properties (1)–(3); we need to show that it is maximal with those properties. Suppose for contradiction that some $\delta \notin \Theta_{n+1}^+$ could be adjoined to Θ_{n+1}^+ preserving (1)–(3). Then, in particular, δ would be admissible in Θ_n^+, and because $\delta \notin \Theta_{n+1}^+$, it must be that $\delta \notin \Gamma^+$.

On the other hand, because $\Theta_n^+ \subseteq \Gamma^+$, we have that δ is admissible in Γ^+; and because $(\Gamma^* - \Gamma^-) \subseteq (\Theta_n^* - \Theta_n^-)$, we have that δ cannot be preempted in $(\Gamma^* - \Gamma^-)$ (not being preempted in the larger set). But because we assumed $(\Gamma^+, \Gamma^-, \Gamma^*)$ to be an extension, we would have $\delta \in \Gamma^+$, contradicting the conclusion reached at the end of the previous paragraph.

We conclude that Θ_{n+1}^+ is a maximal subset of Δ satisfying (1)–(3). Now, by defining Θ_{n+1}^- and Θ_{n+1}^* in the obvious way, we obtain that the sequence thus obtained conforms to the construction of Theorem 3.4.4. The limit $(\Theta^+, \Theta^-, \Theta^*)$ is therefore an extension and, by construction,

$$(\Theta^+, \Theta^-, \Theta^*) \leq (\Gamma^+, \Gamma^-, \Gamma^*).$$

Minimality of $(\Gamma^+, \Gamma^-, \Gamma^*)$ gives the desired conclusion. ■

4

Defeasible Consequence Relations

This chapter deals with the definition of a relation of defeasible consequence based on the notion of general extension introduced in Chap. 3, as well as some alternative developments. The basic notion of defeasible consequence is introduced in Section 4.1: The main result is that the relation so defined satisfies Gabbay's three desiderata (Theorem 4.1.6). Section 4.2 presents alternative developments, namely: (1) the important case of seminormal default theories (which turn out to have a unique minimal general extension); (2) defeasible consequence as based on extensions that are nonminimal ("optimal" in the sense of Manna and Shamir); and (3) a variant of general extensions that, at the cost of a slight complication, avoids certain somewhat counterintuitive results. In Section 4.3 we draw conclusions and comparisons to other approaches, and in Section 4.4 we sketch how to give a "transfinite" version of the present approach. Proofs of selected theorems can be found in Section 4.5.

4.1 DEFEASIBLE CONSEQUENCE

As we set out to define a relation of defeasible consequence, it will be convenient to introduce an abbreviated notation for extensions: We will use boldface uppercase Greek letters to stand for triples of sets of defaults, as in $\mathbf{\Gamma} = (\Gamma^+, \Gamma^-, \Gamma^*)$. Similarly, given sequences of sets of defaults Γ_n^+, Γ_n^-, and Γ_n^*, we write $\mathbf{\Gamma}_n$ for $(\Gamma_n^+, \Gamma_n^-, \Gamma_n^*)$. If $\mathbf{\Gamma}_n$ is a sequence of sets of defaults ($n \geq 0$), we write $\lim \mathbf{\Gamma}_n$ for the unique $\mathbf{\Gamma} = (\Gamma^+, \Gamma^-, \Gamma^*)$ such that

$$\Gamma^+ = \bigcup_{n\geq 0} \Gamma_n^+,$$
$$\Gamma^- = \bigcup_{n\geq 0} \Gamma_n^-,$$
$$\Gamma^* = \bigcap_{n\geq 0} \Gamma_n^*.$$

Finally, to simplify notation still further, we write just $W + \varphi$ in place of $W \cup \{\varphi\}$.

Now we take up a property of general extensions closely related to cumulativity. Section 4.5 contains a proof of the following theorem.

THEOREM 4.1.1 *Let* (W, Δ) *be a default theory, and suppose* $\mathbf{\Gamma}$ *is a 3-tuple of subsets of* Δ *such that* $C(\Gamma^+) \models_W \varphi$. *Then* $\mathbf{\Gamma}$ *is a general extension for* (W, Δ) *if and only if it is a general extension for* $(W + \varphi, \Delta)$.

The preceding result shows that any extension $(\Gamma^+, \Gamma^-, \Gamma^*)$ remains such when we adjoin one of its consequences to the world description. This by itself does not tell us anything about extensions other than $(\Gamma^+, \Gamma^-, \Gamma^*)$. In particular, it does not establish cumulativity. Recall that cumulativity is the property that if φ is supported in every extension of a default theory (W, Δ), then the theory $(W + \varphi, \Delta)$ has the same extensions as (W, Δ). Cumulativity fails for general extensions just as for classical extensions (see Example 3.5.1). But the fact that general extensions have a nontrivial (i.e., nonflat) algebraic structure still allows us to define a cautiously monotonic relation of defeasible consequence. It is perhaps worth recalling that Example 3.5.1 is precisely Makinson's counterexample to Cautious Monotony for defeasible consequence based on classical extensions.

This is perhaps the right time to look again at one particular approach that shares with the theory of general extensions a certain *nonlocal* approach to preemption: this is the Constrained Default Logic of Delgrande et al. (1994). In Constrained Default Logic the consistency check on the justification of a default is not carried out for each default individually, but globally, making sure that in any extension the set comprising the justifications of the triggered defaults is consistent with the set of conclusions of triggered defaults. This has some pleasant consequences, especially as regards the handling of disjunctive information. We further discuss Constrained Default Logic in Subsection 4.3.1, but for now we record here that although Constrained Default Logic is both semimonotonic and orthogonal, Cumulativity fails for it just as it does in the theory of general extensions. A comparison of general extensions and Constrained Default

	Cumulat.	Semimonot.	Orthog.	Caut'ly Monot.
GenExt	No	No	No	Yes
ConDL	No	Yes	Yes	?

FIGURE 4.1. Comparison of general extensions (Gen Ext) and Constrained Default Logic (Con DL).

Logic is summarized in Fig. 4.1 (we don't know whether Constrained Default Logic satisfies Cautious Monotony).

For the purposes of this section, in the light of the characterization of minimal extensions given in Theorem 3.4.5, we refer to a sequence $\mathbf{\Gamma}_n$ as a *construction sequence* for a theory (W, Δ), if the following conditions are met:

1. $\Gamma_0^+ = \Gamma_0^- = \emptyset$;
2. Γ_0^* is the set of all defaults from Δ whose prerequisite is W-consistent;
3. Γ_{n+1}^+ is a maximal set of defaults such that
   a. $C(\Gamma_n^+ \cup \Gamma_{n+1}^+)$ is consistent,
   b. every $\delta \in \Gamma_{n+1}^+$ is admissible in Γ_n^+,
   c. no default $\delta \in \Gamma_{n+1}^+$ is preempted in $\Gamma_n^* - \Gamma_n^-$;
4. $\Gamma_{n+1}^- = \{\delta : \delta$ preempted or conflicted in $\Gamma_{n+1}^+\}$;
5. $\Gamma_{n+1}^* = \{\delta : C(\Gamma_{n+1}^+) \not\models_W \neg P(\delta)\}$.

We are now ready to give the official definition of defeasible consequence based on the notion of general extension. We define $\mid\sim$ skeptically by taking the intersection of all (minimal) extensions of the theory, i.e, all extensions that can be obtained by an inductive process of the same kind as the one given in the proof of Theorem 3.4.4.

DEFINITION 4.1.1 *Let (W, Δ) be a default theory. Then for any sentence φ, we say that φ is a* defeasible consequence *of (W, Δ), written as $(W, \Delta) \mid\sim \varphi$, if and only if $C(\Gamma^+) \models_W \varphi$, whenever $\mathbf{\Gamma}$ is a $\leq$-minimal extension of (W, Δ).*

What follows then is an immediate consequence of the definitions and Theorem 3.4.4.

THEOREM 4.1.2 *If $(W, \Delta) \mid\sim \varphi$, then for every construction sequence $\mathbf{\Gamma}_0, \mathbf{\Gamma}_1, \ldots,$ there is n such that $C(\Gamma_n^+) \models_W \varphi$.*

We need to verify that $\mid\sim$ is indeed a well-behaved notion of defeasible consequence, in that it satisfies the conditions of Reflexivity,

Cautious Monotony, and Cut. However, such properties were formulated for a relation $\mid\!\sim$ that holds between sentences, whereas the relation previously defined holds between a default theory and a sentence. Accordingly, we need to reformulate such properties in order to take this fact into account:

1. *Reflexivity* : If $\varphi \in W$ then $(W, \Delta) \mid\!\sim \varphi$;
2. *Cautious Monotony* : $\dfrac{(W, \Delta) \mid\!\sim \varphi,\ (W, \Delta) \mid\!\sim \psi}{(W + \varphi, \Delta) \mid\!\sim \psi}$;
3. *Cut* : $\dfrac{(W, \Delta) \mid\!\sim \varphi,\ (W + \varphi, \Delta) \mid\!\sim \psi}{(W, \Delta) \mid\!\sim \psi}$.

The next two theorems together entail that the preceding properties hold for $\mid\!\sim$.

THEOREM 4.1.3 *Suppose* $(W, \Delta) \mid\!\sim \varphi$ *and let* $\mathbf{\Gamma} = \lim \mathbf{\Gamma}_n$ *be a minimal extension for* $(W + \varphi, \Delta)$. *Then there is a minimal extension* $\mathbf{\Theta} = \lim \mathbf{\Theta}_n$ *for* (W, Δ) *such that* $\mathbf{\Gamma} \leq \mathbf{\Theta}$.

This theorem shows that although minimal extensions for (W, Δ) start out more slowly than extensions for $(W + \varphi, \Delta)$, they eventually "catch up." Of course, the proof of Theorem 4.1.3 depends crucially on the minimality of $\mathbf{\Gamma}$.

THEOREM 4.1.4 *Suppose* $(W, \Delta) \mid\!\sim \varphi$ *and let* $\mathbf{\Gamma} = \lim \mathbf{\Gamma}_n$ *be a minimal extension for* $(W + \varphi, \Delta)$. *Then there is a minimal extension* $\mathbf{\Theta} = \lim \mathbf{\Theta}_n$ *for* (W, Δ) *such that* $\mathbf{\Theta} \leq \mathbf{\Gamma}$.

Theorem 4.1.4 is the converse of the previous Theorem 4.1.3 and, in a sense, the crucial result in this chapter. Theorem 4.1.4 is what is needed to establish cautious monotonicity for $\mid\!\sim$ and, characteristically, this is where the mathematics itself becomes interesting. The reader is referred to Section 4.5 for the proof: Here we note only that, in a somewhat unexpected twist, the proof of Theorem 4.1.4 makes use of Theorem 4.1.3.

Notice that if $(W, \Delta) \mid\!\sim \varphi$ then (by the two preceding theorems) every minimal extension $\mathbf{\Gamma}$ for $(W + \varphi)$ is between two minimal extensions $\mathbf{\Theta}$ and $\mathbf{\Pi}$ for (W, Δ), i.e., $\mathbf{\Theta} \leq \mathbf{\Gamma} \leq \mathbf{\Pi}$. By minimality we have $\mathbf{\Theta} = \mathbf{\Gamma} = \mathbf{\Pi}$, so $\mathbf{\Gamma}$ is an extension for (W, Δ). This shows that every minimal extension for $(W + \varphi, \Delta)$ is a minimal extension for (W, Δ).

The converse also holds: Let $\mathbf{\Gamma}$ be a minimal extension for (W, Δ). Then, by Theorem 4.1.1, $\mathbf{\Gamma}$ is an extension for $(W + \varphi, \Delta)$: We need to see that it is minimal. If not, then there is an extension $\mathbf{\Theta}$ for $(W + \varphi, \Delta)$ that is properly below $\mathbf{\Gamma}$: $\mathbf{\Theta} < \mathbf{\Gamma}$. By Theorem 4.1.4 there is an extension $\mathbf{\Pi}$ for (W, Δ) such that $\mathbf{\Pi} \leq \mathbf{\Theta}$. By transitivity, we get $\mathbf{\Pi} < \mathbf{\Gamma}$, contradicting the hypothesis that $\mathbf{\Gamma}$ is a minimal extension for (W, Δ). This shows that every minimal extension for (W, Δ) is also a minimal extension for $(W + \varphi, \Delta)$. Thus we have established the following theorem.

THEOREM 4.1.5 *Suppose* $(W, \Delta) \mathrel{|\!\sim} \varphi$. *Then* (W, Δ) *and* $(W + \varphi, \Delta)$ *have exactly the same* minimal *extensions.*

The reader should take notice of the crucial role that the notion of *minimality* plays in the proof of Theorem 4.1.5 (which, in turn, establishes Cautious Monotony). The connection between minimality and Cautious Monotony seems to be more than a mere accident. For instance, one finds the same kind of connection in the work of Kraus et al. (1990) and their system **C** (which was already mentioned in Section 1.2: The connection here is with the minimality of states in the preferential ordering $\prec$). It is also clear now why failure of cumulativity does not preclude the relation of defeasible consequence (as based on general extensions) from being cautiously monotonic: by adding the "theorem" φ back into the world description W we do get new extensions, but such new extensions are not minimal. This fact is subsequently exploited in the proof of Theorem 4.1.6.

THEOREM 4.1.6 *The relation* $\mathrel{|\!\sim}$ *satisfies the properties of Cut, Reflexivity, and Cautious Monotony.*

4.2 ALTERNATIVE DEVELOPMENTS

In this section we consider possible alternative developments of the theory of general extensions. First, we take up the case of seminormal default theories. Such theories are shown to have a unique minimal extension, so that defeasible consequence appears to have a particularly simple definition for seminormal default theories. Second, we identify certain particular, "optimal" extensions of seminormal default theories (based on the lattice-theoretic approach of Manna and Shamir, 1986), and argue that in some cases such extensions might provide more intuitive results, especially in the light of Example 3.5.2. Finally, we identify a variant of the notion of general extension that, although slightly more complicated than the one of Section 3.4, seems also to take care of the problematic case of Example 3.5.2.

4.2.1 Seminormal Theories

Recall from Section 3.2 that a default is seminormal if its justification implies the conclusion, e.g., because the justification contains the conclusion as a conjunct. A theory all of whose defaults are seminormal is also called seminormal. Seminormal theories were first introduced by Reiter and Criscuolo (1981) as a generalization of the case of *normal* theories (those in which the justification is equivalent to the conclusion), which are known always to have a classical extension (in Reiter's sense). Seminormal default theories were developed to handle conflicts among defaults and to block certain unwanted instances of transitivity of implication. In this sense, they form a natural and well-behaved class, which seems sufficient for most purposes in knowledge representation. The downside is that, in contrast to the case of *normal* default theories, seminormal default theories might fail to have a classical extension (see Reiter and Criscuolo, 1981).

But things are quite different when seminormal theories are used in connection with the notion of general extension, for then not only are these theories guaranteed to have an extension (as they are with several other notions of extension), but they are guaranteed to have a uniquely minimal one. This crucial algebraic fact, established in the next theorem, allows for a particularly simple deterministic iterative process for the construction of such a minimal extension.

THEOREM 4.2.1 *Every seminormal default theory has a unique minimal extension.*

The reader is referred to Section 4.5 for a complete proof, but it is worth noting here the basic conceptual fact that the proof exploits. We have seen that the proof of the fact that every default theory has a general extension proceeds by a genuine inductive process. However, such a process is nondeterministic. This is particularly easy to see in the case of categorical (prerequisite-free) default theories. In such a case, we begin by putting $\Gamma_0^+ = \Gamma_0^- = \emptyset$, and for the inductive step we choose as Γ_{n+1}^+ a maximal set of defaults all having the following properties: (1) the conclusions of defaults in $\Gamma_n^+ \cup \Gamma_{n+1}^+$ form a consistent set; and (2) no default in Γ_{n+1}^+ is preempted in $\Delta - \Gamma_n^-$ (the other set Γ_{n+1}^- is determined by the choice of Γ_{n+1}^+). (See the proof of Theorem 3.2.3 for the details.)

But now we are dealing with a seminormal default theory. In particular, no *seminormal* default can be conflicted (relative to any set of defaults) without being already preempted (relative to that set of defaults). For if

$C(\Gamma)$ implies $\neg C(\delta)$ for some default δ, then it also implies $\neg P(\delta)$ because, in turn, $P(\delta)$ implies $C(\delta)$ (all these implications can be taken to be relative to the world description W). It follows that the preceding nondeterministic definition can be replaced with a deterministic one, letting Γ^+_{n+1} be the set of *all* defaults that are not preempted in $\Delta - \Gamma^-_n$. By the previous remark, this ensures consistency of the set of conclusions as well. In turn, the existence of a unique minimal extension allows for the direct definition of defeasible consequence for seminormal default theories by putting $(W, \Delta) \mathrel{|\!\sim} \phi$ if and only if $C(\Gamma^+) \models_W \phi$, where $\mathbf{\Gamma} = (\Gamma^+, \Gamma^-, \Gamma^*)$ is the unique minimal extension of (W, Δ). This notion agrees with the one previously given in Definition 4.1.1: If a default theory (W, Δ) has a *least* extension Γ, then for any φ,

$$(W, \Delta) \mathrel{|\!\sim} \varphi \iff C(\Gamma^+) \models_W \varphi.$$

This definition not only still satisfies Gabbay's requirement, but it also appears to vindicate the view, put forward by Horty et al. (1990), that a cautious theory of defeasible inheritance should be *directly skeptical*, in that it defines the set of skeptically acceptable conclusions in a direct way, and not by means of a detour through the intersection of all credulous extensions. As we have seen in Section 1.5, such a view has been challenged by Makinson and Schlechta (1991), but now it appears that at least in the case of seminormal default theories such a directly skeptical approach is indeed feasible. Indeed, we can obtain the set of the skeptically acceptable conclusions warranted by (W, Δ) by simply giving the (deterministic) inductive process of the proof of Theorem 3.4.4, as amended in Theorem 4.2.1.

Moreover, it is possible to piggyback a relation of defeasible consequence for arbitrary default theories on the relation $\mathrel{|\!\sim}$ previously defined, restricted to seminormal ones. In other words, for each set Δ of defaults let $\mathsf{SN}(\Delta)$ be the result of replacing each default

$$\frac{\alpha : \beta}{\gamma} \quad \text{with} \quad \frac{\alpha : \beta \wedge \gamma}{\gamma}.$$

Then we could define for an arbitrary default theory (W, Δ), $(W, \Delta) \mathrel{|\!\sim}^{\mathsf{SN}} \varphi$ iff $[W, \mathsf{SN}(\Delta)] \mathrel{|\!\sim} \varphi$. Indeed, the switch from a default theory to its seminormalized version is a natural and well-motivated move, advocated for instance by both Reiter and Criscuolo (1981) and Delgrande et al. (1994, p. 193) – in fact, the latter argue that it is reasonable to replace arbitrary defaults by categorical, seminormal ones.

4.2.2 Optimal Extensions

We noticed, in discussing Example 3.5.2, that some of the problems with the minimal extension of the theory discussed there could be obviated by stepping up to a nonminimal extension. The problem, of course, is that in general there will be several nonminimal extensions to choose from, and there seems to be no principled way to single one out as privileged.

Again, the case of seminormal default theories provides an important setting in which to address the problem. In fact, in such a case, we can identify, following Manna and Shamir (1986), certain extensions as the "optimal" fixpoints of a certain monotonic operator. We begin by introducing such an operator, along with the necessary auxiliary notions. In this section, given a default theory (W, Δ) we refer to arbitrary triples $\mathbf{\Gamma} = (\Gamma^+, \Gamma^-, \Gamma^*)$ of subsets of Δ as *pseudoextensions*. If a pseudoextension $\mathbf{\Gamma}$ satisfies the additional condition that Γ^+ and Γ^- are disjoint, and $\Gamma^+ \subseteq \Gamma^*$, then it is called a *potential extension*. Moreover, as we did also elsewhere, we assume that the background world description W is consistent – otherwise, there is only one extension $(\emptyset, \Delta, \emptyset)$.

Definition 4.2.1 *Given a seminormal default theory* (W, Δ)*, define an operator* $\boldsymbol{\tau}$ *over the class of pseudoextensions for* (W, Δ) *by putting* $\boldsymbol{\tau}(\mathbf{\Gamma}) = \mathbf{\Theta}$*, where*

$$\begin{aligned}\Theta^+ &= \{\delta : \delta \textit{ is admissible in } \Gamma^+ \textit{ and} \\ &\qquad \delta \textit{ is not preempted in } \Gamma^* - \Gamma^- \textit{ relative to } W\}, \\ \Theta^- &= \{\delta : \delta \textit{ preempted or conflicted in } \Theta^+ \textit{ relative to } W\}, \\ \Theta^* &= \{\delta : P(\delta) \textit{ is consistent with } C(\Theta^+) \textit{ relative to } W\}.\end{aligned}$$

It is immediate to verify that $\boldsymbol{\tau}$ is a monotone operator over the class of the pseudoextensions, i.e., if $\mathbf{\Gamma} \leq \mathbf{\Theta}$ then $\boldsymbol{\tau}(\mathbf{\Gamma}) \leq \boldsymbol{\tau}(\mathbf{\Theta})$. We are going to be interested in the fixed points of $\boldsymbol{\tau}$, i.e., pseudoextensions $\mathbf{\Theta}$ such that $\boldsymbol{\tau}(\mathbf{\Theta}) = \mathbf{\Theta}$. Of course, these fixed points need not be extensions in the usual sense: $\boldsymbol{\tau}$ will have many fixed points $\mathbf{\Theta}$ in which Θ^+ and Θ^- will not be disjoint or in which Θ^+ is not a subset of $\Theta^* - \Theta^-$.

This is where we can use the conceptual machinery introduced by Manna and Shamir (1986). Following their terminology, say that a collection C of pseudoextensions is $\leq$*-consistent* if and only if any two pseudoextensions in C have an upper bound in C: i.e., if $\mathbf{\Gamma}$ and $\mathbf{\Theta}$ are in C then there is a pseudoextension $\mathbf{\Pi}$ such that both $\mathbf{\Gamma} \leq \mathbf{\Pi}$ and $\mathbf{\Theta} \leq \mathbf{\Pi}$. Now let

$\mathcal{PE}$ be the set of all *potential* extensions for (W, Δ). It is then possible to verify that

C1: Any $\leq$-consistent subset of $\mathcal{PE}$ has a least upper bound in $\mathcal{PE}$; and
C2: Any nonempty subset of $\mathcal{PE}$ has a greatest lower bound in $\mathcal{PE}$.

DEFINITION 4.2.2 *Let* $\mathbf{\Gamma}$ *be a fixed point of* τ. *Then* $\mathbf{\Gamma}$ *is called* intrinsic *if for any other fixed point* $\mathbf{\Theta}$ *of* τ *the set* $\{\mathbf{\Gamma}, \mathbf{\Theta}\}$ *is* $\leq$*-consistent. The largest intrinsic fixed point of* τ *is called* optimal. *(See Manna and Shamir, 1986, p. 419.)*

THEOREM 4.2.2 *Let* (W, Δ) *be a default theory and* $\mathcal{PE}$ *be the collection of all potential extensions for* (W, Δ). *Then* $\mathcal{PE}$ *contains an optimal fixed point of* τ.

This is just an application of Theorem 3 of Manna and Shamir (1986, p. 417) using conditions **C1** and **C2**.

THEOREM 4.2.3 *Let* (W, Δ) *be a seminormal theory, and suppose* $\mathbf{\Gamma}$ *is a potential extension that is also a fixed point of* τ. *Then* $\mathbf{\Gamma}$ *is a general extension for* (W, Δ).

It follows from the theorem that every seminormal default theory has a unique "optimal" extension. It is easy to check that the nonminimal extension of the theory of Example 3.5.2 is also the unique optimal extension. Figure 4.2 depicts the positive parts of the two extensions of the theory: the least one, for which no defaults are triggered, and the maximal (and therefore also optimal) one, which triggers δ_1. It appears then that optimal extensions, at least in the case of seminormal default theories, provide a feasible alternative to minimal extensions, an alternative, moreover, that seems capable of avoiding any residual degree of counterintuitiveness of minimal extensions.

FIGURE 4.2. Nonminimal extensions.

Again, it is immediate to use optimal extensions to define a relation of defeasible consequence: Just say that $(W, \Delta) \mid\!\sim \varphi$ if and only if $C(\Gamma^+) \models_W \varphi$, where Γ is the optimal extension of (W, Δ). However, it is not known at this stage if optimal extensions can support a relation of defeasible consequence that also satisfies Cautious Monotonicity. The crucial role played by minimality in establishing Cautious Monotonicity when $\mid\!\sim$ is defined by use of minimal extensions makes it at least not obvious that this property carries over to optimal extensions.

4.2.3 Circumspect Extensions

The problematic case given in Example 3.5.2 suggests a variant of our notion of extension. Recall that the problem with Example 3.5.2 appeared to be that we allowed defaults in Γ^* even though they were not even potentially admissible, in that their prerequisite was not a logical consequence of conclusions of defaults in Δ. Now, as remarked, this is in keeping with the intuition that potentially preempted defaults should not be triggered, independently of whether the preempting defaults are themselves potentially admissible.

However, the further intuition that defaults that are not even potentially admissible should be ruled out can be incorporated in a rather straightforward way into our definition of extension. This has the cost of complicating the proofs somewhat, but appears at the same time to take care of whatever residual degree of counterintuitiveness the notion of extension might have.

In what follows we consider general extensions with this new twist, which we refer to as "circumspect extensions."

Definition 4.2.3 *A* circumspect extension *for a default theory* (W, Δ) *is a triple* $(\Gamma^+, \Gamma^-, \Gamma^*)$ *of sets of defaults from* Δ*, such that*

- Γ^+ *and* Γ^- *are disjoint;*
- *the following two fixpoint equations are simultaneously satisfied:*

$$\Gamma^+ = \{\delta : \delta \textit{ admissible in } \Gamma^+, \delta \textit{ not conflicted in } \Gamma^+, \textit{and}$$
$$\delta \textit{ not preempted in } \Gamma^* - \Gamma^-\};$$
$$\Gamma^- = \{\delta : \delta \textit{ conflicted or preempted in } \Gamma^+\} \cup$$
$$\{\delta : \neg P(\delta) \textit{ is } W\textit{-consistent with } C(\Gamma^* - \Gamma^-)\};$$

- $\mathsf{Ad}(\Gamma^+) \subseteq \Gamma^* \subseteq \{\delta : P(\delta) \textit{ is } W\textit{-consistent with } C(\Gamma^+)\}$.

Thus Γ^+ is the set of all defaults admissible in Γ^+ but neither conflicted in Γ^+ nor preempted in $(\Gamma^* - \Gamma^-)$. *On the other hand, Γ^- is the set of all defaults either* (1) *conflicted or preempted in Γ^+ or* (2) *whose prerequisite is not implied by the potentially triggered defaults. Finally, Γ^* is a set of defaults containing all default admissible in Γ^+ and whose prerequisite is consistent with $C(\Gamma^+)$.*

It is now incumbent us to show that the theory of general extension carries over to circumspect extensions. However, there is no obstacle to doing so, except perhaps the cost of the added complication in proofs resulting from the new twist in the definition. As evidence of this, the next theorem shows that circumspect extensions always exist (the proof can be found in Section 4.5).

THEOREM 4.2.4 *Let* (W, Δ) *be a default theory. Then there is an iteratively definable circumspect extension* $\boldsymbol{\Theta}$ *for* (W, Δ).

To see how this works, let us take up again Example 3.5.2. Recall that we had a default theory (W, Δ), with W empty and Δ comprising the defaults

$$\delta_1 = \frac{: p}{p}, \qquad \delta_2 = \frac{q : \neg p}{\neg p}.$$

Because $C(\Delta) \not\models_W q$, the default δ_2 is not even potentially admissible in Δ and therefore not even potentially admissible in any $\Theta^* \subseteq \Delta$. It follows that if $\boldsymbol{\Theta}$ is a circumspect extension for the theory, then $\delta_2 \in \Theta^-$. Thus δ_1 is no longer preempted in $\Theta^* - \Theta^-$, and therefore $\delta_1 \in \Theta^+$, as one would expect. Indeed, it is easy to see that $(\delta_1, \delta_2, \delta_1)$ is the only circumspect extension of the theory. However, as we know, it is not the only general extension. Therefore it appears that the net effect of switching to circumspect extensions is to eliminate some of the minimal extensions. In this sense, this solution to the problem presented by Example 3.5.2 is analogous to the one in which one steps up to a nonminimal (e.g., optimal) general extension.

4.3 CONCLUSIONS AND COMPARISONS

As we have seen, the present approach has two main technical fallouts, namely the fact that extensions are always guaranteed to exist and the fact that general extensions allow for the definition of a well-behaved relation of defeasible consequence. In this section, we draw some comparisons with other approaches to defeasible reasoning by means of default rules and look at the question of how the approach based on general extensions fares with the respect to issues of Sections 1.5 and 1.6.

4.3.1 Existence of Extensions

The former problem, that of the existence of extensions, has long been considered one of the basic problems of default logic. This problem has been analyzed from the point of view of computational complexity in Etherington (1987), Dimopoulos and Magirou (1994), Kautz and Selman (1991), and Selman (1990). In these works, broad classes of default theories have been singled out, for which extensions (in Reiter's sense) can be proved to exist. One such class comprises the *ordered default theories* of Etherington (1987) and Selman (1990). An even more general approach is advocated in Papadimitriou and Sideri (1994), in which it is shown that even default theories always have extensions. Indeed, there are reasons to believe that, within the framework of Reiter's approach, the result of Papadimitriou and Sideri (1994) is optimal and cannot be improved on.

Several proposals have been put forward for somewhat different notions of extensions for default theories to solve this problem. As mentioned, extensions always exist according to the notions of extensions proposed by Lukaszewicz (1988), Brewka (1991), and Delgrande et al. (1994). However, these proposals are characterized by an essentially different interpretation of the role of the justification of a default. According to the original intuition of Reiter (1980) (which is shared by the present approach) the justification of a default is to be interpreted as a mere *consistency condition* on the "triggering" of the default. This allows, for instance, for the simultaneous triggering of defaults having mutually inconsistent justifications (provided, of course, that the conclusions are not in turn also inconsistent). This sometimes leads to counterintuitive results as in the "broken arms" example of Poole (1988) (discussed in Section 3.3).

There is a second interpretation, championed for instance by Delgrande et al. (1994), that rather views the justification of a default as a "working hypothesis," whose truth (as opposed to its mere consistency) is to be assumed until and unless information to the contrary becomes available. This intuition quite naturally leads to the so-called *commitment to the justifications*, one of whose consequences, for instance, is that defaults with mutually inconsistent justifications cannot be simultaneously fired, in spite of the fact that their conclusions might be consistent. In other words, the intuition behind this interpretation is that we cannot entertain inconsistent working hypotheses and that therefore the truth of these working hypotheses should be assumed not only in each individual firing of a default, but also across firings of different defaults.

It is important to observe that extensions (in any of the senses available in the literature) for default theories can rarely be constructed (when they exist) by means of a *cumulative process*, of the sort in which defaults are successively assessed for some kind of property that can guarantee their belonging to the extension being constructed. On the contrary, in most cases, we first have to "guess" a set Θ of defaults and then check that it does indeed satisfy the equation defining extensions.

This seems to be connected, at an intuitive level, with the fact that the notions of extension employed by Reiter and many others are intrinsically two-valued, in the sense that such extensions contain the consequences of a maximal set of defaults whose justifications are consistent with the extension itself. This means, among other things, that the triggering of a default can be prevented only if its justification or conclusion is explicitly refuted. Let us informally refer to an approach to Default Logic as *bold* if it shares this feature that any admissible default not explicitly preempted or conflicted is triggered. The formal counterpart to this idea is that *one* set of defaults is used to accomplish a *twofold* task, namely the specification of which defaults are triggered and which defaults are preempted.

On the other hand, the notions of extension proposed in this book (for both nonmonotonic networks and default theories) are essentially three-valued, a feature derived from the analogy to Kripke's (Kripke, 1975) approach to the theory of truth. In Default Logic, when general extensions are used, a default having a sentence φ as its justification (and whose conclusion is otherwise consistent) is prevented from being triggered if and only if $\neg\varphi$ is not explicitly rejected. On the other hand, with Reiter's notion of extension, a default having a sentence φ as its justification can be prevented from being triggered if and only if $\neg\varphi$ is explicitly asserted.

A further trait differentiating Reiter's notion of extensions from general extensions is that the latter can in many cases be obtained as the limit of a genuine inductive construction. By this is meant that general extensions can be constructed "from below" in stages. With Reiter's notion, such a construction from below is possible only when the theory is *semimonotonic*, a desirable feature that is not always easy to enforce (see Delgrande et al., 1994, for a discussion of the desirability of semimonotonicity and the ways it can be achieved).

The intuitions at the basis of the notion of general extension here proposed are quite different also from those underlying the approach of Lukaszewicz, Brewka, or Delgrande et al., and this in spite of the superficial resemblances. If Reiter's notion of extension is to be generalized in such a way as to allow every default theory to have an extension, our proposal generalizes in a different direction from the one previously mentioned, a direction giving rise, as we have already seen, to quite different mathematical properties (some of these properties are summarized in Fig. 4.1).

This is perhaps the right place to point out that an approach somewhat similar to ours has been developed for logic programming. A three-valued or well-founded semantics for logic programs has been developed by van Gelder et al. (1991), and has found what is perhaps the most general formulation in the notion of the *stable model* of Fitting (1990). And, of course, similarities between logic programs and default theories have long been known (see Przymusiński, 1991, who applies such three-valued models to defeasible representational formalisms, although not explicitly to Default Logic).

However, the approach of this chapter differs from the preceding approach in its implementation of the three-valued intuition. In particular, in our approach, the underlying logic is thoroughly classical. The three-valued intuition is cashed out, not by adopting a three-valued underlying logical framework, but rather by identifying extensions for Default Logic, not with sets of sentences or defaults, but with *pairs* (or, in the general case, *triples*) of such sets. This allows for a more concrete representation of extensions, whose properties can then be more easily investigated. Moreover, the present approach gives rise, when applied to arbitrary default theories (as opposed to categorical or prerequisite-free ones), to both *monotonic* and *antimonotonic* processes that in general do not seem to arise in the well-founded approaches to logic programming, except in the general approach of Fitting (1990).

4.3.2 Defeasible Consequence – Again

We now come to the second technical fallout of the notion of general extension, namely the definition of a well-behaved relation of defeasible consequence. By using the notions of extension for Default Logic available in the literature, there are several ways in which we can define, given a default theory having multiple incomparable extensions, a relation $\mid\!\sim$ of defeasible consequence:

1. We can decide to be *credulous* and say that $(W, \Delta) \mid\!\sim \varphi$ precisely when φ follows from (the set of conclusions of defaults in) *some* extension of (W, Δ);
2. We can arbitrarily pick an extension Γ among the many possible, and decide that $(W, \Delta) \mid\!\sim \varphi$ just in case φ follows from (conclusions of defaults in) Γ;
3. We can be *skeptical* and say that $(W, \Delta) \mid\!\sim \varphi$ just in case φ follows from (conclusions of defaults in) *all* extensions of (W, Δ).

All three alternatives have drawbacks. Alternative 2 is not acceptable unless we have a principled way to make such a choice of Γ. Alternative 1 can lead us sometimes to endorse contradictory statements. Alternative 3 is the one that best resonates with certain intuitions about defeasible reasoning, e.g., the fact that defeasible reasoners should be cautious in drawing their inferences (see Horty et al., 1990, for a general argument in favor of skepticism in defeasible reasoning), but it seems to be going about it the wrong way. It does not appear to be an appropriate "implementation" of skepticism to generate *all* possible extensions of a theory and then take the intersection, a point made also by Makinson and Schlechta (1991), although in a different way. The contrast here is with a *directly skeptical* approach that would generate the set of conclusions that are skeptically acceptable *without* going through the detour of first generating all credulous extensions. Moreover, if feasibility of computation is an issue at all, the *intersection-of-extensions* approach is by far the least resource oriented. To clinch matters, as shown by Makinson (1994), the intersection-of-extensions approach, which uses Reiter's notion, simply does not give rise to a cautiously monotonic relation of defeasible consequence for Default Logic.

The cautiously three-valued character of general extensions is responsible for the fact that at least in some important cases there is a natural definition of a notion of logical consequence for default theories. In particular, as we have seen, in the case of seminormal default theories, there is a privileged (unique minimal) general extension that can be used to define a notion of defeasible consequence. [It is also worth remarking that at least in the case of seminormal default theories the least extension can be obtained by means of a genuine inductive construction – in this being similar to, e.g., "prerequisite-free constrained default logic" (PfConDL) of Delgrande et al. (1994), which also allows for a similar construction, but using quite different underlying intuitions.] In contrast, there is no privileged extension in any of the senses of Reiter (1980), Lukaszewicz (1988), Brewka (1991), or Delgrande et al. (1994). In fact, extensions are always taken to be maximal, meaning that any two distinct extensions are $\subseteq$-incomparable, if not outright mutually inconsistent.

4.3.3 Floating Conclusions, Conflicts, and Modularity

We saw in Section 1.5 that an issue arises in connection with so-called floating conclusions, in which a given conclusion is supported in every extension of a theory, but in different ways, in each theory, so that one is left with the choice between accepting a conclusion without a supporting argument or losing the conclusion itself. Horty has convincingly argued that floating conclusions should not be endorsed as a matter of course. Indeed, any directly skeptical approach to defeasible reasoning is bound to lose at least some floating conclusion, a trait that was originally characterized as one of the approach's shortcomings. We can now see that this trait, far from being a shortcoming, is a desirable feature, as it leaves open the possibility of endorsing or rejecting the floating conclusion. The theory of defeasible consequence based on the notion of general extension, to the extent that it falls within the directly skeptical approach, is also likewise immune to floating conclusions, as one can easily see, for example, by formalizing the network of Fig. 1.3 (Section 1.5) as a default theory.

A further issue that we need to consider in this respect arises in connection with the nature and locality of conflicts (see Section 1.6). We first observe that the approach based on general extensions is essentially *modular*. The definition of general extension is built on top of the classical notion of consequence $\models$, which, as is well known, does not handle conflicts in a "local" fashion at all. But the definition could be given equally well (indeed, verbatim) by some other relation, e.g., the four-valued semantics of Belnap (1977), which is close enough to a "relevant" approach to localize conflicts, at least to an extent.

On the general extension approach, any failures of locality can be traced back to $\models$. Consider for instance the default theory (see Section 1.6):

$$\frac{: B}{B}, \quad \frac{: A}{\neg A}.$$

Here, things work out as expected: Under most accounts of conflict, the two sentences A and $\neg A$ are in conflict; hence the second default is self-defeating and cannot be fired. The way general extensions are constructed, this conflict is localized. The second default is potentially preempted, but nothing prevents triggering the first one. Accordingly, the first default is triggered in the unique extension of the theory (this is essentially the same as Example 3.3.5 in Section 3.3).

Now consider the default theory of Brewka and Gottlob (1997) (see Section 1.6):

$$\frac{: B}{B}, \quad \frac{: A}{A}, \quad \frac{: \neg A}{\neg A}.$$

Here, it is the last two defaults that are mutually conflicting and hence potentially preempted. Consequently, neither one is fired in the least extension of the theory, as one would expect (there are, of course nonminimal extensions in which one is triggered and the other one defeated). However, the first default is also *not fired* in the minimal extension, because it is preempted by the conclusions of the last two (potentially active) defaults. But notice that the first default if defeated only because $A, \neg A \models \neg B$, and in turn this depends crucially on a relevance failure of $\models$, and the resulting explosive nature of conflict. In particular, it does not depend on the way general extensions are constructed from the underlying consequence relation.

This point can also be made somewhat differently. Consider the following example:

$$\frac{: B}{B}, \quad \frac{: C}{\neg D}, \quad \frac{: D}{\neg C}.$$

Again, the last two defaults are mutually defeating, but have nothing to do with the first one. This example is then structurally the same as the preceding one, except that now things are set up in such a way as to avoid the explosive, nonlocal nature of conflict. In particular, it is no longer the case that the conclusions of the last two defaults preempt the first: $\neg D, \neg C \not\models \neg B$. It follows that, as one would want, in the minimal extension of the theory, the first two defaults are defeated, but the first one is triggered.

4.4 INFINITELY MANY DEFAULTS

Here it is indicated briefly how to extend the present approach to default theories (W, Δ) comprising infinitely many defaults. For simplicity, we consider only categorical default theories, which form a natural and well-behaved class. As observed in Section 3.2 (and in particular in the proof of Theorem 3.2.3), the hypothesis that Δ is finite is used only once, in order to obtain a certain combinatorial fact. The general case in which Δ is infinite can be handled as follows.

DEFINITION 4.4.1 *Let* (W, Δ) *be a categorical default theory, and* $\delta, \gamma \in \Delta$. *Then* δ *is* below γ, *denoted as* $\delta \prec \gamma$, *iff there are defaults* $\delta_1, \ldots, \delta_n$ *such that*

$$C(\delta_1), \ldots, C(\delta_n), C(\delta) \models_W \neg J(\gamma),$$

but for no proper *subset* $\Gamma_0 \subset \{\delta_1, \ldots, \delta_n, \delta\}$ *we have* $C(\Gamma_0) \models_W \neg J(\gamma)$.

In other words, $\delta \prec \gamma$ iff δ is part of a *minimal* tuple of defaults preempting γ.

DEFINITION 4.4.2 *Let* (W, Δ) *be a categorical default theory. For each* $\delta \in \Delta$, *put*

$$\mathsf{rk}(\delta) = \sup\{\mathsf{rk}(\gamma) + 1 : \gamma \in \Delta \;\&\; \gamma \prec \delta\},$$

if such a sup *exists, and* $\mathsf{rk}(\delta) = \infty$ *otherwise [in which case we say that* $\mathsf{rk}(\delta)$ *is undefined].*

Similarly, put $\mathsf{rk}(\Delta) = \sup\{\mathsf{rk}(\delta) : \mathsf{rk}(\delta)$ *is defined*$\}$.

It is then possible to carry out the construction of Theorem 3.2.3 transfinitely through the ordinal $\mathsf{rk}(\Delta)$, appropriately taking unions at limit stages. It is not difficult to see that this yields a general extension. The crucial step is that if $\delta_1, \ldots, \delta_n$ is a (minimal) n-tuple preempting δ then $\mathsf{rk}(\delta_i) < \mathsf{rk}(\delta)$ for every i, so we can apply an appropriate inductive hypothesis to conclude that such an n-tuple must have a member in Γ^-.

As an added bonus, we have the following: For any default theory (W, Δ), let the *well-founded part* of Δ be $\Theta = \{\delta : \mathsf{rk}(\delta)$ is defined$\}$. Then it can be verified that any general extension for (W, Δ) is a *classical* extension for (W, Θ).

4.5 PROOFS OF SELECTED THEOREMS

THEOREM 4.1.1 Let (W, Δ) be a default theory, and suppose $(\Gamma^+, \Gamma^-, \Gamma^*)$ is a 3-tuple of subsets of Δ such that $C(\Gamma^+) \models_W \varphi$. Then $(\Gamma^+, \Gamma^-, \Gamma^*)$ is a general extension for (W, Δ) if and only if it is a general extension for $(W \cup \{\varphi\}, \Delta)$.

Proof. We do the case for a categorical default theory, the general case being similar. Because $C(\Gamma^+) \models_W \varphi$, any default δ is conflicted in Γ^+ relative to W if and only if it is conflicted in Γ^+ relative to $W \cup \{\varphi\}$. Thus, in order to establish the theorem it suffices to establish that δ is preempted in $\Delta - \Gamma^-$ relative to W if and only if δ is so preempted relative to $W \cup \{\varphi\}$.

One direction is immediate: If δ is preempted in $\Delta - \Gamma^-$ relative to W then it is still so preempted relative to $W \cup \{\varphi\}$, by monotonicity of classical logic.

For the converse, assume that δ is preempted in $\Delta - \Gamma^-$ relative to $W \cup \{\varphi\}$. Then

$$C(\Delta - \Gamma^-) \models_{W \cup \{\varphi\}} \neg J(\delta); \tag{4.1}$$

But by hypothesis, $C(\Gamma^+) \models_W \varphi$, and by disjointness of Γ^+ and Γ^-, also $\Gamma^+ \subseteq (\Delta - \Gamma^-)$. By monotonicity of classical logic, $C(\Delta - \Gamma^-) \models_W \varphi$, whence by expression (4.1) and Cut (for classical logic), also $C(\Delta - \Gamma^-) \models_W \neg J(\delta)$, as desired. ■

THEOREM 4.1.3 Suppose $(W, \Delta) \mid\!\sim \varphi$ and let $\mathbf{\Gamma} = \lim \mathbf{\Gamma}_n$ be a minimal extension for $(W + \varphi, \Delta)$. Then there is a minimal extension $\mathbf{\Theta} = \lim \mathbf{\Theta}_n$ for (W, Δ) such that $\mathbf{\Gamma} \leq \mathbf{\Theta}$.

Proof. Because $(W, \Delta) \mid\!\sim \varphi$ and every default theory has a minimal extension, let $\mathbf{\Pi} = \lim \mathbf{\Pi}_n$ be a minimal extension for (W, Δ) such that $C(\Pi^+) \models_W \varphi$. Let $k > 0$ be an integer such that already $C(\Pi_k^+) \models_W \varphi$.

We are going to define a construction sequence $\mathbf{\Theta}_n$. For $m \leq k$ we put $\mathbf{\Theta}_m = \mathbf{\Pi}_m$. For $k + n$ (where $n > 0$) we put

$$\begin{aligned}
\Theta_{k+n}^+ &= \text{a maximal subset of } \Delta \text{ extending } \Gamma_n^+, \text{ such that} \\
&\quad \text{(A) } C(\Theta_{k+n-1}^+ \cup \Theta_{k+n}^+) \text{ is } W\text{-consistent,} \\
&\quad \text{(B) every } \delta \in \Theta_{k+n}^+ \text{ is admissible in } \Theta_{k+n-1}^+, \\
&\quad \text{(C) no } \delta \in \Theta_{k+n}^+ \text{ is preempted in } \Theta_{k+n-1}^* - \Theta_{k+n-1}^-; \\
\Theta_{k+n}^- &= \{\delta : \delta \text{ preempted or conflicted in } \Theta_{k+n}^+ \text{ relative to } W\}; \\
\Theta_{k+n}^* &= \{\delta : P(\delta) \text{ is consistent with } C(\Theta_{k+n}^+) \text{ relative to } W\}.
\end{aligned}$$

Because any maximal subset of Δ extending Γ_n^+ and having properties (A), (B), and (C) is also a maximal subset of Δ having properties (A), (B), and (C), we obtain immediately that $\mathbf{\Theta}_n$ is a construction sequence and that $\mathbf{\Theta} = \lim \Theta_n$ is a minimal extension for (W, Δ).

So we need to show $\mathbf{\Gamma} \leq \mathbf{\Theta}$; in turn, it suffices to show that $\mathbf{\Gamma}_n \leq \mathbf{\Theta}_{k+n}$. This we do by induction on n.

Case $n = 0$. Because we have $\Gamma_0^+ = \Gamma_0^- = \emptyset$, all we need to show is that $\Theta_{k+0}^* \subseteq \Gamma_0^* = \{\delta : P(\delta) \text{ is } W + \varphi \text{ consistent}\}$. Now,

$$\Theta_{k+0}^* = \{\delta : P(\delta) \text{ is } W + \varphi\text{-consistent with } C(\Theta_{k+0}^+)\}.$$

Because by construction $C(\Theta_{k+0}^+) \models_W \varphi$, we have that if $P(\delta)$ is $W + \varphi$ inconsistent then it is also W-inconsistent with $C(\Theta_{k+0}^+)$. So if $\delta \notin \Gamma_0^*$ then also $\delta \notin \Theta_{k+0}^*$, as desired. This shows that $\mathbf{\Gamma}_0 \leq \mathbf{\Theta}_{k+0}$.

Case $n + 1$. Assume that $\mathbf{\Gamma}_n \leq \mathbf{\Theta}_{k+n}$ in order to show that $\mathbf{\Gamma}_{n+1} \leq \mathbf{\Theta}_{k+n+1}$. By construction, Θ_{k+n+1}^+ extends Γ_{n+1}^+. To prove that $\Gamma_{n+1}^- \subseteq \Theta_{n+1}^-$, recall that, by definition,

$$\Theta_{k+n+1}^- = \{\delta : \delta \text{ preempted or conflicted in } \Theta_{k+n+1}^+ \text{ relative to } W\},$$
$$\Gamma_{n+1}^- = \{\delta : \delta \text{ preempted or conflicted in } \Gamma_{n+1}^+ \text{ relative to } W + \varphi\}.$$

Now, if δ is preempted or conflicted in Γ_{n+1}^+ relative to $W + \varphi$, because $C(\Theta_{k+n+1}^+) \models_W \varphi$ and $\Gamma_{n+1}^+ \subseteq \Theta_{k+n+1}^+$, δ is also preempted or, respectively, conflicted in Θ_{k+n+1}^+ relative to W. So $\Gamma_{n+1}^- \subseteq \Theta_{n+1}^-$.

Finally, ad $\Theta_{k+n+1}^* \subseteq \Gamma_{n+1}^*$. We have

$$\Theta_{k+n+1}^* = \{\delta : P(\delta) \text{ is } W\text{-consistent with } \Theta_{k+n+1}^+\},$$
$$\Gamma_{n+1}^* = \{\delta : P(\delta) \text{ is } W + \varphi\text{-consistent with } \Gamma_{n+1}^+\}.$$

Suppose $\delta \notin \Gamma_{n+1}^*$; then $C(\Gamma_{n+1}^+) \models_{W+\varphi} \neg P(\delta)$. Because $\Gamma_{n+1}^+ \subseteq \Theta_{k+n+1}^+$ and $C(\Theta_{k+n+1}^+) \models_W \varphi$, also $C(\Theta_{k+n+1}^+) \models_W \neg P(\delta)$, whence $\delta \notin \Theta_{k+n+1}^*$, as desired.

This concludes the induction, showing $\mathbf{\Gamma} \leq \mathbf{\Theta}$. ∎

Theorem 4.1.4 Suppose $(W, \Delta) \mid\!\sim \varphi$ and let $\mathbf{\Gamma} = \lim \mathbf{\Gamma}_n$ be a minimal extension for $(W + \varphi, \Delta)$. Then there is a minimal extension $\mathbf{\Theta} = \lim \mathbf{\Theta}_n$ for (W, Δ) such that $\mathbf{\Theta} \leq \mathbf{\Gamma}$.

Proof. Because $\mathbf{\Gamma}$ is a minimal extension for $(W + \varphi, \Delta)$ and $(W, \Delta) \mid\!\sim \varphi$, we know from Theorem 4.1.3 that there is a minimal extension $\mathbf{\Pi}$ for (W, Δ) above $\mathbf{\Gamma}$ (in the $\leq$ ordering). Pick such an extension, which will remain fixed for the rest of the proof. We will show that there is an extension for (W, Δ) below $\mathbf{\Gamma}$: Of course it will follow that there is a *minimal* extension for (W, Δ) below $\mathbf{\Gamma}$.

Such an extension $\mathbf{\Theta}$ that is below $\mathbf{\Gamma}$ (in the ordering $\leq$) will be defined as the limit of an increasing sequence, i.e., $\mathbf{\Theta} = \lim \mathbf{\Theta}_n$. For $n = 0$ we put $\Theta_0^+ = \Theta_0^- = \emptyset$, and $\Theta_0^* = \{\delta : P(\delta)$ is W consistent$\}$.

For the inductive step:

$$
\begin{aligned}
\Theta_{n+1}^+ = {} & \text{a maximal subset of } \Pi^+ \text{ such that} \\
& \text{(A) } C(\Theta_n^+ \cup \Theta_{n+1}^+) \text{ is } W\text{-consistent,} \\
& \text{(B) every } \delta \text{ in } \Theta_{n+1}^+ \text{ is admissible in } \Theta_n^+ \text{ relative to } W, \\
& \text{(C) no } \delta \in \Theta_{n+1}^+ \text{ is preempted in } \Theta_n^* - \Theta_n^-, \text{ relative to } W.
\end{aligned}
$$

Having done this, we define Θ_{n+1}^- and Θ_{n+1}^* as usual as the set of defaults preempted or conflicted in Θ_{n+1}^+ relative to W, and, respectively, as the set of defaults whose prerequisite is consistent with Θ_{n+1}^+ relative to W.

So we can put $\mathbf{\Theta} = \lim \mathbf{\Theta}_n$. We need to establish the following:

1. The sequence $\mathbf{\Theta}_n$ is increasing in the ordering $\leq$,
2. $\mathbf{\Theta}$ is an extension for (W, Δ),
3. $\mathbf{\Theta} \leq \mathbf{\Gamma}$.

The proof for the first two items is precisely similar to the corresponding items in the proof of Theorem 3.4.4, and we skip it. We concentrate on the last, crucial, item.

To show that $\mathbf{\Theta} \leq \mathbf{\Gamma}$, it suffices to show $\mathbf{\Theta}_n \leq \mathbf{\Gamma}$ by induction on n. The case for $n = 0$ is easy. We have $\Theta_0^+ = \emptyset \subseteq \Gamma^+$, and similarly for Θ_0^-. Now $\{\delta : P(\delta)$ is W-consistent$\} = \Theta_0^*$: if $P(\delta)$ is W-inconsistent then it is also $W + \varphi$ inconsistent with $C(\Gamma^+)$, so if $\delta \notin \Theta_0^*$ then $\delta \notin \Gamma^*$, i.e., $\Gamma^* \subseteq \Theta_0^*$.

Now the inductive step for $n + 1$. We assume that $\delta \in \Theta_{n+1}^+$ in order to show that $\delta \in \Gamma^+$; in turn, we establish this by proving that δ is admissible in Γ^+ relative to $W + \varphi$, not conflicted in Γ^+ relative to $W + \varphi$ and not preempted in $\Gamma^* - \Gamma^-$ relative to $W + \varphi$:

1. δ is admissible in Θ_n^+ relative to W and hence (because $\Theta_n^+ \subseteq \Gamma^+$) also admissible in Γ^+ relative to $W + \varphi$.
2. By inductive hypothesis, we have $(\Gamma^* - \Gamma^-) \subseteq (\Theta_n^* - \Theta_n^-)$. Suppose for contradiction that δ is preempted in $(\Gamma^* - \Gamma^-)$ relative to $W + \varphi$. Then

$$(\Gamma^* - \Gamma^-) \models_{W+\varphi} \neg J(\delta).$$

We show $(\Gamma^* - \Gamma^-) \models_W \varphi$, whence

$$(\Gamma^* - \Gamma^-) \models_W \neg J(\delta),$$

which in turn gives (by monotonicity of $\models_W$) $(\Theta_n^* - \Theta_n^-) \models_W \neg J(\delta)$, which is impossible.

To prove $(\Gamma^* - \Gamma^-) \models_W \varphi$: We have that $\mathbf{\Pi}$ is a minimal extension for (W, Δ): Because $(W, \Delta) \mid\!\sim \varphi$, we have $C(\Pi^+) \models_W \varphi$. Because $\mathbf{\Pi}$ is an extension, all defaults in Π^+ are admissible in Π^+ and hence they belong to Π^*; because Π^+ and Π^- are disjoint, we have $\Pi^+ \subseteq (\Pi^* - \Pi^-)$. It follows that $(\Pi^* - \Pi^-) \models_W \varphi$, and because $\mathbf{\Gamma} \leq \mathbf{\Pi}$ also $(\Gamma^* - \Gamma^-) \models_W \varphi$, as desired.

3. Finally, we show that δ is not conflicted in Γ^+ relative to $W + \varphi$. If δ were so conflicted, then $C(\Gamma^+) \models_{W+\varphi} \neg C(\delta)$, and because $\Gamma^+ \subseteq \Pi^+$, also $C(\Pi^+) \models_{W+\varphi} \neg C(\delta)$.

   But now, as before, $C(\Pi^+) \models_W \varphi$, so $C(\Pi^+) \models_W \neg C(\delta)$. In other words, δ is conflicted in Π^+ relative to W, which is impossible given that $\delta \in \Theta_{n+1}^+ \subseteq \Pi^+$ and $\mathbf{\Pi}$ is an extension for (W, Δ).

As mentioned, this gives $\Theta_{n+1}^+ \subseteq \Gamma^+$. From this, we can obtain (by using the fact that $S \models_W \psi$ implies $S \models_{W+\varphi} \psi$) that $\Theta_{n+1}^- \subseteq \Gamma^-$ and $\Gamma^* \subseteq \Theta_{n+1}^*$. In turn, this gives $\mathbf{\Theta} \leq \mathbf{\Gamma}$. ■

THEOREM 4.1.6 The relation $\mid\!\sim$ satisfies the properties of Cut, Reflexivity, and Cautious Monotony.

Proof. We take up the different properties in turn. As we will see, Theorem 4.1.5 will play a crucial role.

1. *Reflexivity*: We need to show that if $\varphi \in W$ then $(W, \Delta) \mid\!\sim \varphi$. This follows immediately: If $(\Gamma^+, \Gamma^-, \Gamma^*)$ is any extension (minimal or otherwise) of the theory, and $\varphi \in W$, then in particular $C(\Gamma^+) \models_W \varphi$, so that $(W, \Delta) \mid\!\sim \varphi$.
2. *Cautious Monotonicity*: We need to show

$$\frac{(W, \Delta) \mid\!\sim \varphi, \quad (W, \Delta) \mid\!\sim \psi}{(W + \varphi, \Delta) \mid\!\sim \psi}.$$

   Because $(W, \Delta) \mid\!\sim \varphi$ and $(W, \Delta) \mid\!\sim \psi$, then for every $\leq$-minimal extension $\mathbf{\Gamma} = (\Gamma^+, \Gamma^-, \Gamma^*)$ of (W, Δ) we have $C(\Gamma^+) \models_W \varphi \wedge \psi$. By Theorem 4.1.5, each such extension is also a $\leq$-minimal extension of $(W + \varphi, \Delta)$, and conversely. By monotonicity of classical logic, $C(\Gamma^+) \models_{W+\varphi} \psi$, whence $(W + \varphi, \Delta) \mid\!\sim \psi$, as desired.
3. *Cut*: We need to show

$$\frac{(W, \Delta) \mid\!\sim \varphi, \quad (W + \varphi, \Delta) \mid\!\sim \psi}{(W, \Delta) \mid\!\sim \psi.}.$$

Let $\mathbf{\Gamma} = (\Gamma^+, \Gamma^-, \Gamma^*)$ be $\leq$-minimal extension of (W, Δ): We need to show $C(\Gamma^+) \models_W \psi$. By Theorem 4.1.1 and the first premise of Cut, $\mathbf{\Gamma}$ is also a $\leq$-minimal extension of $(W + \varphi, \Delta)$. By the second premise of Cut, $C(\Gamma^+) \models_{W+\varphi} \psi$. But also $C(\Gamma^+) \models_W \varphi$: By using Cut for classical logic, we have $C(\Gamma^+) \models_W \psi$, as desired. ■

THEOREM 4.2.1 Every seminormal default theory has a unique minimal extension.

Proof. We do the case for (W, Δ) categorical, the general case being similar. We define, as in the proof of Theorem 3.2.3, a sequence of pairs of sets (Γ_n^+, Γ_n^-) of defaults, and we begin by putting $\Gamma_0^+ = \Gamma_0^- = \emptyset$. For the inductive step we put

$$\Gamma_{n+1}^+ = \{\delta : \delta \text{ not preempted in } \Delta - \Gamma_n^-\},$$
$$\Gamma_{n+1}^- = \{\delta : \delta \text{ preempted or conflicted in } \Gamma_{n+1}^+\}.$$

We assume that the sequence (Γ_n^+, Γ_n^-) coincides with the one in the proof of Theorem 3.2.3 up to stage n and show that this must be the case also at stage $n + 1$. Observe that the inductive hypothesis yields, in particular, that $C(\Gamma_n^+)$ is W-consistent.

Now Γ_{n+1}^+ has the property of being a maximal set of defaults not preempted in $\Delta - \Gamma_n^-$ (being the set of *all* such defaults). So if we can show that it also has the further property that $C(\Gamma_n^+ \cup \Gamma_{n+1}^+)$ is W-consistent, it will follow that Γ_{n+1}^+ is a maximal set of defaults having the *two* mentioned properties, and we will have recovered the construction given in the proof of Theorem 3.2.3.

So we show that $C(\Gamma_n^+ \cup \Gamma_{n+1}^+)$ is W-consistent. Suppose for contradiction that this fails. We know that $C(\Gamma_n^+)$ is W-consistent by itself, so that if $C(\Gamma_n^+ \cup \Gamma_{n+1}^+)$ is W-inconsistent, it must be that $\Gamma_{n+1}^+ \neq \emptyset$. We will contradict this last fact, showing that $\Gamma_{n+1}^+ = \emptyset$.

From the hypothesis that $C(\Gamma_n^+ \cup \Gamma_{n+1}^+)$ is W-inconsistent, it follows that there are defaults $\delta_1, \ldots, \delta_k \in \Gamma_{n+1}^+$ such that

$$C(\Gamma_n^+) \models_W \neg[C(\delta_1) \wedge \ldots \wedge C(\delta_k)]. \tag{4.2}$$

Now we show that, for each default δ_i among $\delta_1, \ldots, \delta_k$, we have $\delta_i \in (\Delta - \Gamma_n^-)$. Reasoning by reductio, suppose that $\delta_i \in \Gamma_n^-$. Then δ_i is either conflicted or preempted in Γ_n^+. But δ_i is seminormal, so that it cannot be conflicted without being already preempted. So this implies that δ_i is preempted in Γ_n^+, and because $\Gamma_n^+ \subseteq (\Delta - \Gamma_n^-)$, it follows that δ_i is preempted in $(\Delta - \Gamma_n^-)$. But by definition, this is equivalent to $\delta_i \notin \Gamma_{n+1}^+$, against assumption. We conclude that $\delta_1, \ldots, \delta_k \in (\Delta - \Gamma_n^-)$.

From expression (4.2) we have that

$$C(\delta_1), \ldots, C(\delta_n) \models_W \neg C(\Gamma_n^+),$$

where $\neg C(\Gamma_n^+)$ is to be construed as the negation of the conjunction of conclusions of defaults in Γ_n^+. Because $\delta_1, \ldots, \delta_k \in (\Delta - \Gamma_n^-)$, also

$$C(\Delta - \Gamma_n^-) \models_W \neg C(\Gamma_n^+).$$

But Γ_n^+ and Γ_n^- are disjoint, so that $\Gamma_n^+ \subseteq (\Delta - \Gamma_n^-)$. We conclude that $C(\Delta - \Gamma_n^-)$ must be W-inconsistent, so that any default is preempted in $(\Delta - \Gamma_n^-)$, whence $\Gamma_{n+1}^+ = \emptyset$. This is the contradiction we sought.

We conclude that the construction given here coincides with the one given in the proof of Theorem 3.2.3 in the case of seminormal theories (or Theorem 3.4.4 in the noncategorical case), and therefore the limit of the sequence of pairs of sets of defaults yields a general extension. The process is now deterministic, so this extension is unique. We already know from Theorem 3.4.4 that any extension obtained in this way is minimal. Theorem 3.4.5 gives uniqueness. ■

THEOREM 4.2.3 Let (W, Δ) be a seminormal theory, and suppose $\boldsymbol{\Gamma}$ is a potential extension that is also a fixed point of $\boldsymbol{\tau}$. Then $\boldsymbol{\Gamma}$ is a general extension for (W, Δ).

Proof. Let $\boldsymbol{\Theta}$ be a fixed point of $\boldsymbol{\tau}$. First we observe that, because W is consistent, Θ^+ must also be W-consistent: If not, then any δ is preempted in Θ^+ relative to W, so $\Theta^- = \Delta$, and by disjointness $\Theta^+ = \emptyset$. Therefore, if Θ^+ is W-inconsistent, it must be that W is already inconsistent.

Given that $\boldsymbol{\Theta}$ is a fixed point, we easily verify the equations defining extensions. Similarly, it's immediate to see that if $P(\delta)$ is W-inconsistent then $\delta \notin \Theta^*$. So all that is left to verify is that Θ^* contains all δ admissible in Θ^+: But if $C(\Theta^+) \models_W P(\delta)$ then by W-consistency of Θ^+ also $C(\Theta^+) \not\models_W \neg P(\delta)$, i.e., $P(\delta)$ is W-consistent with Θ^+, whence $\delta \in \Theta^*$, as desired. ■

THEOREM 4.2.4 Let (W, Δ) be a default theory. Then there is an iteratively definable circumspect extension $\boldsymbol{\Theta}$ for (W, Δ).

Proof. We construct a circumspect extension for (W, Δ) iteratively by putting $\Theta_0^+ = \Theta_0^- = \emptyset$ and $\Theta_0^* = \{\delta : P(\delta) \text{ is } W\text{-consistent}\}$.

For the inductive step, as in the proof of Theorem 3.4.4, we put

$$\begin{aligned} \Theta_{n+1}^+ = {} & \text{a maximal set of defaults such that} \\ & \text{(A) } C(\Theta_n^+ \cup \Theta_n^+) \text{ is } W\text{-consistent,} \\ & \text{(B) every } \delta \text{ in } \Theta_{n+1}^+ \text{ is admissible in } \Theta_n^+ \text{ relative to } W, \\ & \text{(C) no } \delta \in \Theta_{n+1}^+ \text{ is preempted in } \Theta_n^* - \Theta_n^-, \text{ relative to } W. \end{aligned}$$

Now for the new twist: We put

$$\begin{aligned} \Theta_{n+1}^- = {} & \{\delta : \delta \text{ conflicted or preempted in } \Theta_{n+1}^+\} \cup \\ & \{\delta : C(\Theta^* - \Theta^-) \not\models_W P(\delta)\}. \end{aligned}$$

On the other hand, the definition of Θ^* is the usual one: Θ^* is the set of defaults δ whose prerequisite $P(\delta)$ is W-consistent with Θ^+.

The first thing to prove is that the sequence we obtain is increasing, i.e., that $\mathbf{\Theta}_n \leq \mathbf{\Theta}_{n+1}$ for each n. This can be shown by induction on n, just as in the proof of Theorem 3.4.4, except at the inductive step, in which we show $\Theta_{n+1}^- \subseteq \Theta_{n+2}^-$. This we do in some detail. So suppose $\delta \in \Theta_{n+1}^-$; to show $\delta \in \Theta_{n+2}^-$, we distinguish two cases.

1. Suppose δ is preempted or conflicted in Θ_n^+; then, by using the inductive hypothesis, we obtain that δ is preempted or conflicted in Θ_{n+1}^+ and hence $\delta \in \Theta_{n+2}^-$.
2. The other case is $C(\Theta_n^* - \Theta_n^-) \not\models_W P(\delta)$. By the inductive hypothesis, $\Theta_{n+1}^* \subseteq \Theta_n^*$ and $\Theta_n^- \subseteq \Theta_{n+1}^-$. It follows that $(\Theta_{n+1}^* - \Theta_{n+1}^-)$ is a subset of $(\Theta_n^* - \Theta_n^-)$, so that $C(\Theta_{n+1}^* - \Theta_{n+1}^-) \not\models_W P(\delta)$. So again we have $\delta \in \Theta_{n+2}^-$.

Finally, we need to show that $\mathbf{\Theta} = \lim \mathbf{\Theta}_n$ is an circumspect extension. The only added complication over the proof of Theorem 3.4.4 is when we check that

$$\text{if } C(\Theta^* - \Theta^-) \not\models_W P(\delta) \text{ then } \delta \in \Theta^-.$$

We can assume that δ is not conflicted or preempted in Θ^+ and hence not in any Θ_n^+, for if it is then we immediately get $\delta \in \Theta^-$. So suppose that $\delta \notin \Theta^-$. Then for each n we have $C(\Theta_n^* - \Theta_n^-) \models_W P(\delta)$. But the sequence $C(\Theta_n^* - \Theta_n^-)$ is *decreasing* in n, so eventually there must be a tuple of defaults $\delta_1, \ldots, \delta_k$ and some n such that

$$\delta_1, \ldots, \delta_k \in \bigcap_{m \geq n} (\Theta_m^* - \Theta_m^-),$$

and, moreover,

$$C(\delta_1), \ldots, C(\delta_k) \models_W P(\delta).$$

Then $\delta_1, \ldots, \delta_k \in (\Theta^* - \Theta^-)$, whence $C(\Theta^* - \Theta^-) \models_W P(\delta)$, as required.

■

Bibliography

A. R. Anderson, N. D. Belnap, and J. M. Dunn, *Entailment: The Logic of Relevance and Necessity*, Princeton University Press, Princeton, NJ, 1975–1992, 2 vols.

G. A. Antonelli, Defeasible inheritance over cyclic networks, *Artificial Intelligence*, 92(1):1–23, 1997.

G. A. Antonelli, A directly cautious theory of defeasible consequence for default logic via the notion of general extension, *Artificial Intelligence*, 109(1–2):71–109, 1999.

G. A. Antonelli, Logic, in L. Floridi, 2004, pp. 263–75.

F. Baader and B. Hollunder, Computing extensions of terminological default theories, in G. Lakemeyer and B. Nebel, editors, *Foundations of Knowledge Representation and Reasoning*, Vol. 810 of Lecture Notes in Computer Science, pp. 35–110, Springer-Verlag, New York and Berlin, 1994.

C. Baral and V. S. Subrahmanian, Duality between alternative semantics of logic programs and nonmonotonic formalisms, in *Proceedings of the First International Workshop on Logic Programming and Nonmonotonic Reasoning*, pp. 69–86, MIT Press, Cambridge, MA, 1991.

N. D. Belnap, A useful four-valued logic, in J. M. Dunn and G. Epstein, editors, *Modern Uses of Multiple-Valued Logic*, Reidel, Dordrecht, The Netherlands, 1977.

G. Boole, *An Investigation of the Laws of Thought*, Walton and Maberly, London, 1854.

E. Börger, E. Grädel, and Y. Gurevich, *The Classical Decision Problem*, Springer-Verlag, Berlin, 1997.

C. Boutilier, *On the Semantics of Stable Inheritance Reasoning*, Tech. Rep., Dept. of Computer Science, University of Toronto, 1989.

R. Brachman and H. Levesque, editors, *Readings in Knowledge Representation*, Morgan Kaufmann, Los Altos, CA, 1985.

S. Brass, On the semantics of supernormal defaults, in *IJCAI-93. Proceedings of the Thirteenth Joint Conference on Artificial Intelligence*, Vol. I, pp. 578–83, Morgan Kaufmann, San Mateo, CA, 1993.

G. Brewka, Cumulative default logic: In defense of nonmonotonic inference rules, *Artificial Intelligence*, 50:183–205, 1991.

G. Brewka and G. Gottlob, Well-founded semantics for default logic, *Fundamenta Informaticæ*, XX:1–16, 1997.

B. F. Chellas, *Modal Logic: An Introduction*, Cambridge University Press, New York, 1980.

A. Church, An unsolvable problem of elementary number theory, *American Journal of Mathematics*, 58:345–363, 1936.

M. J. Cresswell and G. E. Hughes, *A New Introduction to Modal Logic*, Routledge, London, 1995.

R. Dedekind, *Was sind und was sollen die Zahlen?*, Brunswick, 1888.

J. P. Delgrande, T. Schaub, and W. K. Jackson, Alternative approaches to default logic, *Artificial Intelligence*, 70:167–237, 1994.

Y. Dimopoulos and V. Magirou, A graph theoretic approach to default logic, *Journal of Information and Computation*, 112:239–56, 1994.

J. Dix, Default theories of Poole-type and a method for constructing cumulative versions of default logic, in B. Neumann, editor, *Proceedings of the 10th European Conference on Artificial Intelligence (ECAI-92)*, pp. 289–93, Wiley, New York, 1992.

H.-D. Ebbinghaus, J. Flum, and W. Thomas, *Mathematical Logic*, 2nd ed., Springer-Verlag, New York, 1994.

H. Enderton, *A Mathematical Introduction to Logic*, Academic Press, New York, 1972; 2nd ed., 2001, published by Harcourt/Academic Press.

D. Etherington, Formalizing nonmonotonic reasoning systems, *Artificial Intelligence*, 31:41–85, 1987.

D. Etherington and R. Reiter, On inheritance hierarchies with exceptions, in *Proceedings of AAAI–83*, pp. 104–8, American Association for Artificial Intelligence, Menlo Park, CA, 1983.

S. Feferman, Kurt Gödel, Conviction and caution, in *In the Light of Logic*, Oxford University Press, New York, 1998.

M. Fitting, *First-Order Logic and Automated Theorem Proving*, Springer-Verlag, New York, 1990.

M. Fitting, The family of stable models, *Journal of Logic Programming*, 17:197–225, 1993.

L. Floridi, editor, *Blackwell Guide to the Philosophy of Computing and Information*, Blackwell, Oxford, UK, 2004.

G. Frege, *Begriffsschrift, eine der arithmetische nachgebildete Formelsprache des reines Denkens*, L. Nebert, Halle, 1879; English translation in van Heijenoort, 1967.

D. M. Gabbay, Theoretical foundations for nonmonotonic reasoning in expert systems, in K. Apt, editor, *Logics and Models of Concurrent Systems*, pp. 439–59, Springer-Verlag, Berlin, 1985.

D. M. Gabbay, C. J. Hogger, and J. A. Robinson, editors, *Handbook of Logic in Artificial Intelligence and Logic Programming*, Vol. 3, Oxford University Press, Oxford, UK, 1994.

M. L. Ginsberg, editor, *Readings in Nonmonotonic Reasoning*, Morgan Kauffman, Los Altos, CA, 1987.

K. Gödel, Die Vollständigkeit der Axiome des logischen Funktionenkalküls, *Monatshefte für Mathematik und Physik*, 37:349–60, 1930; English translation as "The completeness of the axioms of the functional calculus of logic" in van Heijenoort, 1967, pp. 582–91.

A. Gupta and N. D. Belnap, *The Revision Theory of Truth*, MIT Press, Cambridge, MA, 1993.

P. J. Hayes, The logic of frames, in D. Metzing (ed.), *Frame Conceptions and Text Understanding*, de Gruyter, Berlin 1979, pp. 46–61; reprinted in Brachman and Levesque, 1985, pp. 287–295.

D. Hilbert, Mathematische Probleme, *Göttinger Nachrichten*, 1900, pp. 253–297, also appeared in *Archiv der Mathematik und Physik*, (3) 1 (1901), pp. 44–63 and 213–237.

J. F. Horty, Some direct theories of nonmonotonic inheritance, in D. M. Gabbay et al., 1994, pp. 111–187.

J. F. Horty, Skepticism and floating conclusions, *Artificial Intelligence Journal*, 135:55–72, 2002.

J. F. Horty, R. H. Thomason, and D. S. Touretzky, A skeptical theory of inheritance in nonmonotonic semantic networks, *Artificial Intelligence*, 42:311–48, 1990.

H. Kautz and B. Selman, Hard problems for simple default logic, *Artificial Intelligence Journal*, 49:243–279, 1991.

S. C. Kleene, *Introduction to Metamathematics*, Van Nostrand, Princeton, NJ, 1952.

S. Kraus, D. Lehman, and M. Magidor, Nonmonotonic reasoning, preferential models and cumulative logics, *Artificial Intelligence*, 44(1–2):167–207, 1990.

S. Kripke, Outline of a theory of truth, *The Journal of Philosophy*, 72:690–716, 1975.

D. Lehman and M. Magidor, What does a conditional knowledge base entail?, *Artificial Intelligence*, 55(1):1–60, 1992.

V. Lifschitz, Computing circumscription, in M. L. Ginsberg, 1987, pp. 167–73.

W. Łukaszewicz, Considerations on default logic, *Computational Intelligence*, 4:1–16, 1988.

D. Makinson, General patterns in nonmonotonic reasoning, in D. M. Gabbay et al., 1994, pp. 35–110.

D. Makinson and K. Schlechta, Floating conclusions and zombie paths: Two deep difficulties in the "directly skeptical" approach to defeasible inheritance networks, *Artificial Intelligence*, 48:199–209, 1991.

Z. Manna and A. Shamir, The theoretical aspects of the optimal fixedpoint, *SIAM Journal of Computing*, 5:414–26, 1986.

B. Nebel, *Reasoning and Revision in Hybrid Representation Systems*, Vol. 422 of Lecture Notes in Artificial Intelligence Series, Springer-Verlag, New York, 1990.

C. H. Papadimitriou and M. Sideri, Default theories that always have extensions, *Artificial Intelligence*, 69:347–57, 1994.

D. L. Poole, A logical framework for default reasoning, *Artificial Intelligence*, 36:27–48, 1988.

H. Przymusińska and T. Przymusiński, Stationary default extensions, *Fundamenta Informaticæ*, 21(1–2):76–87, 1994.

T. Przymusiński, Three-valued nonmonotonic formalisms and semantics of logic programs, *Artificial Intelligence*, 49:309–43, 1991.
R. Reiter, A logic for default reasoning, *Artificial Intelligence*, 13:81–132, 1980.
R. Reiter and G. Criscuolo, On interacting defaults, in *Proceedings of the Seventh International Joint Conference on Artificial Intelligence*, pp. 270–6, Vancouver, B.C., Morgan Kaufmann Publishers, Los Altos, CA, 1981.
E. Sandewall, Non-monotonic inference rules for multiple inheritance with exceptions, in *Proceedings of the IEEE*, 74:1345–53, 1986.
K. Schlechta, Directly skeptical inheritance cannot capture the intersection of extensions, *Journal of Logic and Computation*, 3:455–67, 1993.
B. Selman, *Tractable Default Reasoning*, Ph.D. thesis, University of Toronto, 1990.
G. Simonet and R. Ducournau, On Stein's paper: Resolving ambiguity in nonmonotonic inheritance hierarchies, *Artificial Intelligence*, 71:181–93, 1994.
R. Stalnaker, Nonmonotonic consequence relations, *Fundamenta Informaticæ*, 21:7–21, 1994.
L. Stein, Resolving ambiguity in nonmonotonic reasoning, *Artificial Intelligence*, 55:259–310, 1992.
G. Takeuti, *Proof Theory*, 2nd ed., Elsevier North-Holland, New York, 1987.
A. Tarski, Der Wahrheitsbegriff in den formalisierten Sprachen, *Studia Logica*, pp. 261–405, 1935, English translation in A. Tarski, 1956, pp. 152–278.
A. Tarski, *Logic, Semantics, and Metamathematics*, Oxford University Press, Oxford, UK, 1956, edited and translated by J. H. Woodger.
R. H. Thomason, Netl and subsequent path-based inheritance theories, *Computers and Mathematics with Applications*, 23:179–204, 1992.
R. H. Thomason, J. F. Horty, and D. S. Touretzky, A calculus for inheritance in monotonic semantic nets, in Z. W. Ras and M. Zemankove, editors, *Methodologies for Intelligent Systems*, pp. 280–87, Elsevier, New York, 1987.
D. Touretzky, *The Mathematics of Inheritance Systems*, Morgan Kaufmann, Los Altos, CA, 1986.
A. M. Turing, On computable numbers, with an application to the Entscheidungsproblem. *Proceedings of the London Mathematical Society*, Series 2, 42 (1936–37), pp. 230–265.
A. Urquhart, Complexity, in L. Floridi, 2004, pp. 18–26.
A. van Gelder, K. A. Ross, and J. S. Schlipf, The well-founded semantics for general logic programs, *Journal of the Association for Computing Machinary*, 38(3):620–50, 1991.
J. van Heijenoort, editor, *From Frege to Gödel. A Source Book in Mathematical Logic*, Harvard University Press, Cambridge, MA, 1967.
J.-H. You, X. Wang, and L.-Y. Yuan, Compiling defeasible inheritance networks to general logic programs, *Artificial Intelligence*, 113(1–2):247–68, 1999.

Index